Revise for Geography A2 Edexcel Specification B

David Burtenshaw

I am indebted to the years of help from my colleagues at Edexcel, not least Tony, Mike, Sue, Steve, Dulcie, Bob and Frank. Many of their ideas have been used in this guide. I thank them for their inspiration. The errors though are mine alone.

Heinemann Educational Publishers
Halley Court, Jordan Hill, Oxford OX2 8EJ
Part of Harcourt Education
Heinemann is the registered trademark of
Harcourt Education Limited

© David Burtenshaw, 2004

First published 2004

08 07
10 9 8 7 6 5 4

British Library Cataloguing in Publication Data is available
from the British Library on request.

ISBN: 978 0 435101 55 8

Copyright notice

All rights reserved. No part of this publication may be reproduced in any form or by any means (including photocopying or storing it in any medium by electronic means and whether or not transiently or incidentally to some other use of this publication) without the written permission of the copyright owner, except in accordance with the provisions of the Copyright, Designs and Patents Act 1988 or under the terms of a licence issued by the Copyright Licensing Agency, 90 Tottenham Court Road, London W1T 4LP. Applications for the copyright owner's written permission should be addressed to the publisher.

Typeset by Saxon Graphics Ltd, Derby
Original illustrations © Harcourt Education Limited, 2003
Printed in UK by Ashford Colour Press Ltd, Gosport, Hampshire

Acknowledgements

Figures 1.4 (1) PP20, 1.4 (3) PP22, 1.14 (3) PP60, 1.14 (4) PP62, 1.14 (5) PP63, 1.16 (1) PP67, 1.18 (1) PP71 appear courtesy of the *Financial Times*; Figure 1.18 (3) PP72 appears courtesy of the Royal Meteorological Society. Figure 2.2d (2) PP92 appears courtesy of the NOAA.

Every effort has been made to contact copyright holders of material reproduced in this book. Any omissions will be rectified in subsequent printings if notice is given to the publishers.

> ## Websites
>
> On pages where you are asked to go to www.heinemann.co.uk/hotlinks to complete a task or down load information, please insert the code **1552S** at the website.

Contents

Introduction	4

1 Global Challenge (Unit 4, Paper 6474) — 7

A The natural environment

1.1	The global importance of climate and weather	9
1.2	Managing changeable weather globally and in the UK	12
1.3	Seasonal variations in climate	16
1.4.	Climate change	20
1.5	Biomes, ecosystems and the threats to their survival	24
1.6	The degradation and conservation of the world's forests	28
1.7	The deterioration and desertification of the world's grasslands	31
1.8	The importance of marine and coastal ecosystems	34
1.9.	People and the future of ecosystem management	36

B Population and the economy

1.10	Measuring population change	39
1.11	The national and international challenges of population change	43
1.12	International migration	47
1.13	Global economic groupings	52
1.14	Changes in the location of economic activity	58
1.15	The future of the global economy	64
1.16	Addressing the development gap	66
1.17	Sustainable development	69
1.18	Examination questions (Sections A and B)	71

C Cross modular questions

1.19	Dealing with the cross-unit questions (Section C)	73

2 Researching Global Futures: Managing natural environments (Unit 5, 6475/1) — 75

2.1	The Environment and its resources	78
2.2	Living with hazardous environments	81
2.2a	The physical processes causing natural hazards	84
2.2b	The impact of hazards on people, the economy and the environment	86
2.2c	How people respond to and manage natural hazards	89
2.2d	Future issues in living with natural hazards	91
2.3	Pollution and natural environments	94
2.3a	Measuring variations in pollution	96
2.3b	Environmental, social and economic impacts of pollution	99
2.3c	Managing pollution incidents	103
2.3d	Alternative strategies for managing pollution in the future	106
2.4	Wilderness environments	109
2.4a	The significance of wilderness regions	111
2.4b	Pressures on wilderness environments	113
2.4c	Strategies for managing wilderness regions	115
2.4d	Protecting wilderness regions	117
2.5	Examination questions	119

3 Researching Global Futures: Challenges for the human environments (6475/2) — 120

3.1	Development and disparity	123
3.2	Feeding the world's people	125
3.3	Health and welfare	127
3.4	The geography of sport and leisure	129
3.5	Sample essay titles	131

4 Synoptic unit: issues analysis — 133

Answers to exam questions	137
Index	144

Introduction to the guide

How to use this revision guide

This revision guide is designed to help you achieve a good grade in the A2 Edexcel Geography B course. The guide follows the sequence of the specification and covers its three units:

1   Global Challenge (**Unit 4**, 6474)
2   Researching Global Futures: Managing natural environments (**Unit 5**, 6475, paper 1)
    Researching Global Futures: Challenges for the human environments (**Unit 5**, 6475, paper 2)
3   Synoptic Unit: Issues Analysis (**Unit 6**, 6476).

For details of the specification coverage and assessments see page 6.

Each part is divided into sections that cover the specification topics. The guide also has **Quick check questions** and **Reminders** to keep you thinking about what you have read and studied. The Reminders give helpful examination hints and cross-refer you to other topics. The Quick check questions are designed to make you test your recall and understanding of the topics. Wherever possible there is advice based on the experience of examiners.

Part 1 covers the Unit 4 *Global Challenge* (6474). It also covers the cross-modular studies linking the four modules – *The natural environment, Weather and climate, Biomes and ecosystems, Population and the economy*.

Part 2 covers Paper 1 of Unit 5 *Researching Global Futures* (6475/1).

There are revision checklists at the beginning of each of these two parts. At the end of Parts 1 and 2, there are practice questions, which are similar to those set in the examination papers.

Part 3 covers Paper 2 of Unit 5 *Researching Global Futures* (6475/2). This has been treated in a different way because, as students, you will have to submit a research essay. It has been written to give you advice on how to approach this type of research essay and it will give you guidance on the stages of preparation and writing up. There is also a useful set of books and web links to websites to enable you to start your investigations.

Part 4 covers Unit 6 the *Synoptic Unit* (6476) and offers you advice on the best way to approach this synoptic paper. It gives you a study sequence for the critical time between the issuing of the resource materials and the day when you sit the examination. The examination depends largely on your ability to synthesise information across the whole AS and A2 course, as well as to interpret the resource materials issued. Therefore, Part 4 also discusses the best ways to make use of those resources, both before and during the examination.

Jumping to A2

Starting A2 is a big jump. You need to have made the step up as quickly as possible in your work in the autumn term. You have to be able to write at A2 level no matter when you sit your examination. You will not be given a choice if your centre enters you for a paper in the January examinations. Do not let your centre enter you for two papers in January unless you are a genius!

What makes it A2

- You have to write in continuous prose, and under pressure, in all four examinations. In other words you must be able to use a formal essay format for most of your answers. Essays demand that you write in sentences and these are grouped into paragraphs. The paragraphs should be logically ordered and the whole essay must contain both a short introduction to the topic and a brief conclusion that reminds the examiner of what you have written. **Quality of Written Communication** (QWC) gains you up to 10/70 marks for 6476 and 10/60 marks for both 6475/1 and 6475/2. In other words **11.1%** of your marks can be obtained by writing well; that can make the difference of a grade. QWC for 6474 is built into the 80 marks for the paper.

- You have to show a better appreciation and understanding of the interaction between, and complexity of, causes. You must understand the variety and complexity of solutions to issues and problems. There are no simple cause–effect relationships in geography at this level. The interrelationships between people and their environment are very complex.
- You must be able to synthesise, or draw together, information from a variety of resources used in your studies. One source is no longer sufficient.
- Your knowledge must contain both depth and breadth and, wherever possible, include work that you have done on your own.
- You must be prepared to undertake research on your own away from the classroom from a variety of sources and not just the Internet.
- The issues will be complex and controversial. There will be no right or wrong answer to many questions. Therefore you need to appreciate why people hold differing opinions.
- The data you are given in the examination will require more than a descriptive response; it will be a stimulus for your ideas and theories that you have learned.
- 6474 is now being marked online and you have to develop the discipline of writing within the margins. If you have to add sentences after you have written an answer, develop the discipline of adding the detail at the end of the answer on separate sheets. Do use colour pencil despite the request not to use colour. Coloured maps will result in the scripts being marked by hand.
- Finally, you should be able to draw very stylised sketch maps and diagrams. These should aid the answers and not just be for showing locations.

Good luck with your revision: I hope it enables you to achieve the grade that you want.

David Burtenshaw

Summary of modules

Module name	Contents of module	Length of exam	Total number of marks in the exam (% of Advanced GCE*)	Distribution of marks in the exam
Part 1 Global challenge (6474)	**Section A** 1 Global weather and climate 2 Ecosystems **Section B** 3 Population 4 The global economy **Section C** Cross-modular questions	2 hours	84 (15%)	Structured essays using data stimulus materials. Section A and Section B – 2 × 25 marks Section C – 30 marks QWC+ 4 marks
Part 2 Researching global futures: Managing natural environments (6475/1)	ONE of four options: 1 Environments and resources 2 Living with hazardous environments 3 Pollution of natural environments 4 Wilderness environments	1 hour 20 minutes	60 (7.5%)	Formal essay based on research and written under examination conditions. Choice from two essays on a generalisation. Generalisation to be examined issued at least 2 weeks before the examination. See cell below for marks.
Part 3 Researching global futures: Challenges for human environments (6475/2)	ONE of four options: 1 Development and disparity 2 Feeding the world's people 3 Health and welfare 4 Geography of sport and leisure	1500 word essay written as coursework	60 (7.5%)	Coursework report. Selection from a list of titles issued in May for the following year's two examinations. Marks for both halves of 6475 based on: defining the topic – 10, research – 15, understanding – 15, conclusion – 10. QWC – 10.
Part 4 Synoptic assessment – issues analysis (6476)	A synoptic, issues-based paper using pre-released data. Content can range across the whole AS and A2 specification.	2 hours	70 (20%)	Pre-issued resources (about 2 – 3 weeks prior to exam). Exam is a synoptic, issues-based paper of up to 4 questions worth 70 marks with 10 marks for QWC+

* = Remember 50% of the marks for Advanced GCE come from the AS course

+ = Quality of written communication

PART 1

Global Challenge

General principles and strategies

Part 1 Global Challenge (Unit 4, Paper 6474) is very big and will demand that you really do respond as an A2 student, whether you take the paper in January or June. There are three examination sections:

Section A The natural environment

There are two modules:
Weather and climate (1.1–1.4, pages 9–323 in this revision guide)
Biomes and ecosystems (1.5–1.9, pages 24–38 in this revision guide)
NB There are three options in the second module (1.6, 1.7 and 1.8) and your school/college will have selected one for you to study. Your choice of biome (in 1.5) should be related to your option selection.

Section B Population and the economy

There are two modules:
Population (1.10–1.12, pages 39–51 in this revision guide)

The economy (1.13–1.17, pages 52–74 in this revision guide)

NB There are three options in the second module (1.15, 1.16 and 1.17). Your choice might be tutor guided or based on your interests. All options overlap with the economy sections.

Section C Cross-modular questions

The questions in this section will draw on your studies of all the four modules above (1.19, pages 73–74). There is no syllabus for this section. You are expected to be able to integrate ideas, concepts and examples from across your study of Global Challenge. Geography is about the interrelationships between the natural and human environments and this section asks you to do just that.

The key to 6474 is to understand the wide spectrum of challenges faced by the planet and its inhabitants. These are schematised in Figure 1. The revision checklist on page 8 gives both the exam specification and the revision guide numbering.

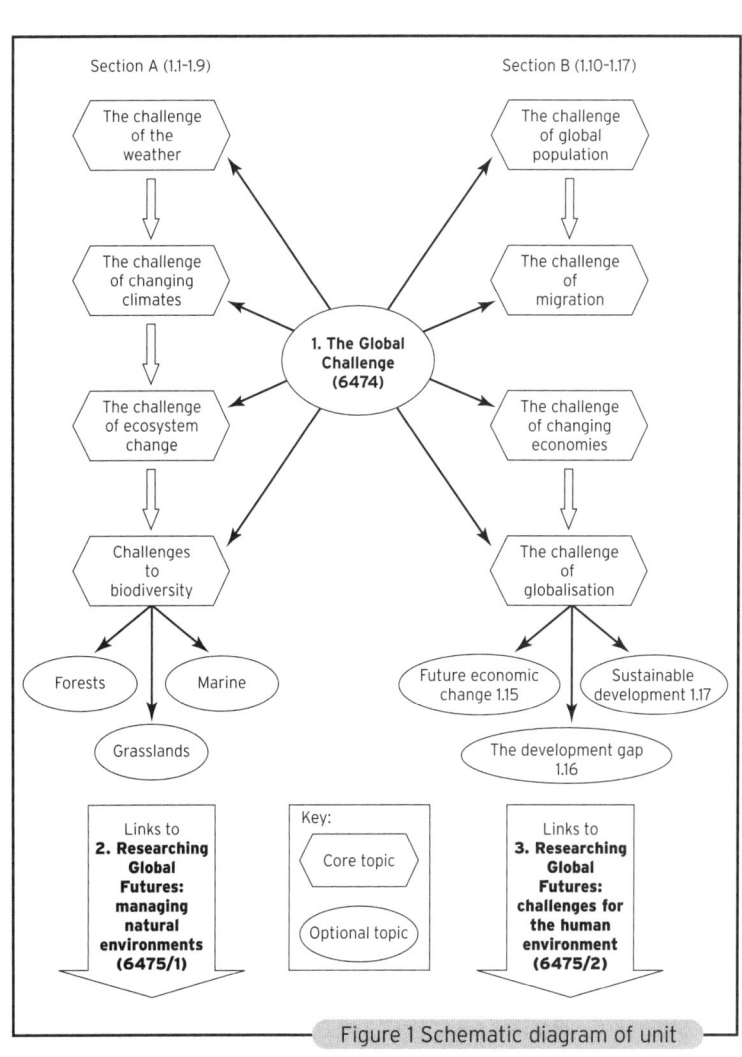

Figure 1 Schematic diagram of unit

Global challenge

Revision checklist

To help you, fill in the checklist below as you progress with your revision, and, when you are certain that you have covered all that is needed, the final column can be filled in as a last check. Do not delude yourself; be honest.

RG = Revision Guide	Enquiry Questions	Have organised notes	Have examples – some of my own	Read textbooks	Know terms	Can draw maps	Have diagrams and models	Can use on cross unit	Completed revision
Section A									
Spec. 4.0 RG 1.1	Weather and climate								
Spec. 4.1 RG 1.2	Mid-latitude areas								
Spec. 4.2 RG 1.3	Seasonal variations								
Spec 4.3 RG 1.4	Climatic change								
Spec 4.4 RG 1.5	Global biomes								
Spec 4.5a RG 1.6	Forest degradation								
Spec 4.5b RG 1.7	Deterioration and desertification								
Spec. 4.5c RG 1.8	Marine ecosystems								
Spec, 4.6 RG 1.9	Ecosystem Management								
Section B									
Spec. 4.7 RG 1.10	Population change								
Spec. 4.8 RG 1.11	Challenges of population change								
Spec. 4.9 RG 1.12	International migration								
Spec. 4.10 RG 1.13	Global economic groupings								
Spec 4.11 RG 1.14	Character and location of global economy								
Spec. 4.12a RG 1.15	Future change								
Spec. 4.12b RG 1.16	Development gap								
Spec. 4.12c RG 1.17	Sustainable development								
Section C Overarching, cross-modular themes									
Biodiversity									
Environmental sustainability									
Poverty									
Economic sustainability									
Conservation versus development									
Environmental degradation and destruction									
Globalisation									

The optional sections, pp28–35 and pp64–70, have been shaded as has the cross-modular section of the examination.

1.1 The global importance of climate and weather

Key questions for this section:

- Why do weather and climate present a global challenge?
- Why is weather prediction important?

You will be revising:

- Weather forecasting and its global challenges.

You have to study at least 'Climate' or 'Ecosystems' because they are the subject matter of Section A in the examination. This is a section where you can use your experience of everyday weather in the UK. 'Weather and climate' can overlap with some work on 'Hazardous environments' (see 2.2, pp 81–92).

What is the importance of weather forecasts?

Weather affects us all and, therefore, we need to know how it might affect our daily lives. Farmers want to know if it will be dry for harvesting; the transport authorities want to know if there will be ice on roads; and the emergency services want to know if flooding might occur. Meteorologists prepare forecasts for different timescales and they are most accurate over a short time period of several days.

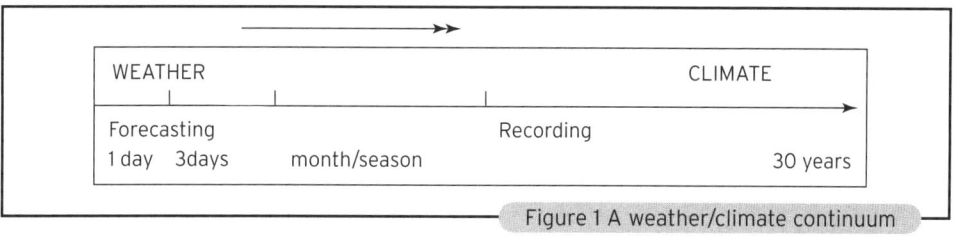

Figure 1 A weather/climate continuum

Forecasts are more important in climates where there is more day-to-day and seasonal variations, e.g. West Europe (see 1.2). Weather affects a range of economic activities in the short, medium and long term as the list below illustrates.

1 Short term forecasts (generally for one or two days ahead) affect:

- **People** – what to wear; driving conditions; what food to buy; imminence of hazards; heating or air conditioning demand depending on season
- **Farmers** – conditions for planting, harvesting, daily tasks, etc.
- **Transport** – problems of ice on power lines, rail tracks, aircraft; fog; heat buckling; emergency services alert; more/less demand
- **Insurance** – alert for rise in claims; more call-outs, e.g. RAC
- **Building work** – dangers from high winds; cold
- **Health** – increasing risks from sunburn, pollen and hay fever, asthma etc.
- **Power suppliers** – changes in demand for air conditioning in summer (e.g. Italy 2003) and heating in winter.

2 Medium-term forecasts (for a period of a week to 10 days) affect:

- **People** – what to buy; recreational activities; preparing for hazards
- **Farmers** – care of crops and livestock; flood risk; irrigation needs
- **Transport** – preparation for extreme events
- **Insurance** – medium term planning and underwriting
- **Building work** – preparation for high winds, e.g. tower crane use
- **Manufacturers** – food producers may alter patterns of output, e.g. cooling fans went out of stock in August 2003
- **Retailers** – ice-cream, beer, iced drinks sales rise in hot summer; more barbecues purchased
- **Health** – increase in pollutants, ozone; asthma, hypothermia risk.

Key concepts

Weather The day-to-day patterns of temperature, precipitation, humidity (visibility), winds.

Meteorology The science of explaining and predicting the processes that influence the weather, i.e. the science of weather.

Climatology The science that documents and explains the area variations in weather, normally over long time-scales.

Climate The average of the weather data over a thirty year period. It is a generalisation and so increasing storminess may be interpreted as a change in climate characteristics.

Reminder

If you are taking the exam in January try to memorise and generalise weather charts from the newspapers that illustrate the weather of the past month. Winter should be good for depressions and blocking anticyclones. If you are entering in June use the charts from the April to May period which are generally good for low pressure systems.

3 **Long-term forecasts (for a period of a month or more and are less accurate) affect:**

- **People** – less important because accuracy is diminished. Long range forecasts have greater margins of error.
- **Farmers** – aids some planning: erosion threats, e.g. wet autumn when fields ploughed for winter wheat
- **Transport** – enables planning of activities and use of labour force
- **Insurance** – enables underwriters to decide on premiums, e.g. driest days for summer weddings = lower premiums
- **Building work** – scheduling of activities
- **Manufacturers and retailers** – better planning of supply to meet demand
- **Tourism** – planning seasonal activities both by companies and individuals.

> **Reminder**
>
> Good weather charts can be found in *The Times*, *The Guardian*, *The Independent*, *The Daily Telegraph* and *The Financial Times*.

Climate is a key factor in the environment. It determines the nature of the ecosystems, the plants and the productivity of an ecosystem. It influences the rate of weathering and erosion. For instance desert environments are altered more rapidly by flash flooding. It is a key to the understanding of biomes (see 1.5, pp 24–27).

Extreme events

Extreme weather costs lives and money. In advanced societies the insurance losses can be huge and premiums go up if costs rise. Some examples to show the value of predicting extreme weather are as follows (N.B. It is nearly always short term):

- Forecasting hurricane from Miami Hurricane Forecast Centre enable precautions, such as evacuation of coastal areas, boarding up and making property safe, to be taken in USA, Central America and the Caribbean e.g. Hurricane Mitch and Hurricane Isobel 2003.
- Predicting the paths of tornados in 'Tornado alley' Oklahoma USA. Precautions can be taken and lives saved.
- Alerting people to the dangers of bush-fires in periods of extreme heat, e.g. Provence in 2003 and New South Wales in December 2002.
- Warning of dangers of flash floods enabling some evacuation.
- Predicting long term extreme weather, such as periods of wet weather that have the cumulative effect of raising the water table, saturating the soil and increasing overland flow.

Why is weather prediction a challenge?

1 If forecasters get it wrong it can have many consequences, e.g. Michael Fish's prediction that there would be no hurricane the evening before the Great Storm of October 1987. In fact it was not a hurricane but the wind hazard was extreme – 15 million trees were destroyed.
2 In a technologically advanced world many activities, especially transport, rely on accurate predictions in order to schedule activities, e.g. salting of roads, changing to snow tyres, preparing a ski slope.
3 The impact of extreme events, such as flooding, hurricanes, tornadoes, is costly both to governments and insurers. It is easier to respond if emergency services are on standby.
4 Levels of predictive capability vary and therefore the challenge is how to assist those countries less able to afford sophisticated prediction.

Figure 2 summaries two responses to weather, the deterministic – we can do little approach; and the technological fix – where great faith is placed in the ability of technology and the brains behind it to solve challenges.

1.1 The global importance of climate and weather

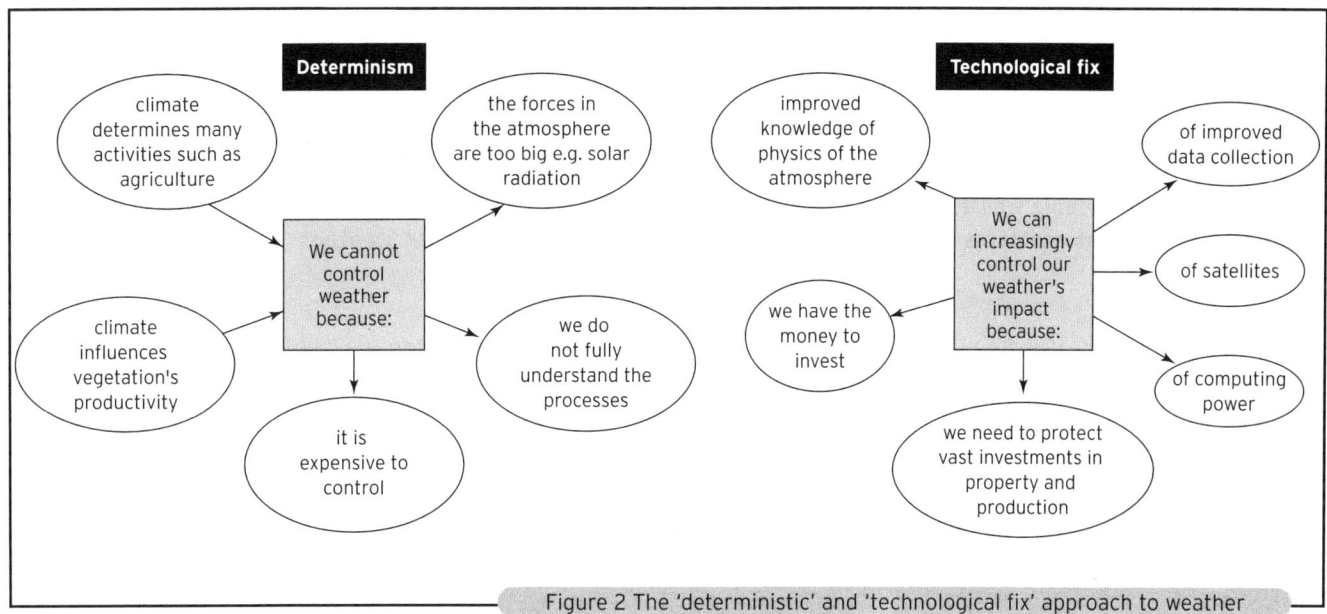

Figure 2 The 'deterministic' and 'technological fix' approach to weather

 Try to explain why neither of the two approaches to weather – deterministic and the technological fix – solve all the problems of the weather.

Reminder

Make sure you know about the influence of climate on the ecosystems that you select in Unit 4.5 (1.5). For Unit 4.5a (1.6) make sure you have data and diagrams for the Equatorial climate for rainforests and Arctic and cool temperate interior for boreal forest. For Unit 4.5b (1.7) you will need to know about the savanna climatic regime and both cool and warm temperate interior climates to help you in your study of grasslands. The links are more varied for coral and other marine/coastal ecosystems.

1.2 Managing changeable weather globally and in the UK

Key questions for this section:

- Why do mid-latitude areas, such as the UK, experience changeable weather?
- What management problems does this changeable weather cause?

You will be revising:

- The global atmospheric system
- Air masses, depressions, anti cyclones
- Their effect on UK weather.

Much of this section is based on your knowledge of UK weather in the context of the global system. Managing changeable weather is an underlying theme.

The global atmospheric system

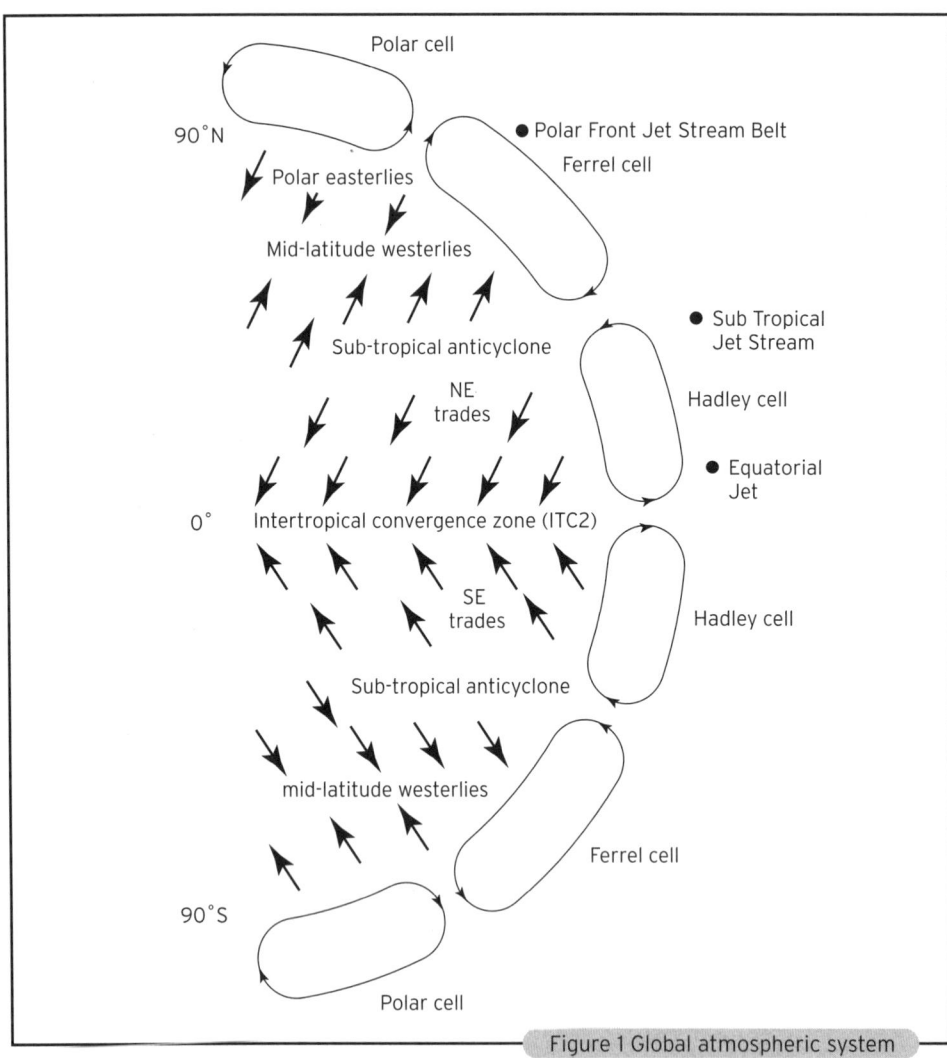

Figure 1 Global atmospheric system

Key concepts

Air masses Air with common characteristics of temperature and humidity acquired from long term surface contact in its area of origin. Each mass is separated from the next by fronts. When air masses move they pick up some of the characteristics of the land or seas that they cross. They help to give variety to UK and West European weather.

Frontogenesis A process where two distinct air masses meet and produce the conditions which lead to the formation of a depression with its own frontal system.

Rossby waves One of the most intriguing natural phenomena, also known as *planetary waves* as they owe their origin to the shape and rotation of the Earth. They are waves in the jet streams in response to the rotation of the Earth, and latitudinal variations in the Coriolis force, that have an important effect on surface weather. Carl-Gustav Rossby was the first to theorise about them in the 1930s.

The global atmospheric system consists of:

- Three circulatory systems: the Hadley, the mid-latitude and the polar
- Main **jet stream** courses above the convergence and divergent zones
- Major high pressure anticyclone zones where air descends, e.g. Azores and Saharan Highs

1.2 Managing changeable weather globally and in the UK

Weather phenomena that affect the UK

Air masses

You need to understand the characteristics and paths of the air masses that affect the UK because they determine the weather conditions. The dominant masses will vary from one season to the next and help to account for the variety of our weather.

Air mass	Origins and track	Characteristics over UK
Arctic maritime (Am)	Polar cell over the cold, frozen Arctic Ocean – tracks south across Arctic Ocean and North Sea	Very cold – some moisture picked up which may fall as snow over east and north-facing coasts and hills. Blizzards in Scotland. Mainly occurs in winter months.
Polar continental (Pc)	Polar cell over land especially Northern and Eastern Europe. Moves west across North Sea	Cold dry air but some moisture picked up across North Sea giving rise to some snow in eastern districts. Mainly occurs in winter months
Polar maritime (Pm)	Originating over North America as Pc but picking up characteristics of the North Atlantic. Tracks south-eastwards to Western Europe.	Moisture picked up over North Atlantic, so therefore unstable. Cumulus clouds and heavy showers. Cool weather to north and west. Affects all seasons.
Returning Polar maritime	Pc air that has moved south into the Atlantic before approaching UK from west or south-west.	Variant of above where air has tracked to warmer south before tracking to UK from south west. Sea fog in summer. Can be similar to Tm below.
Tropical maritime (Tm)	From Azores High. Tracks north east.	Warm moist air is being cooled and therefore stable. Rainfall may be heavy and accentuated by relief (orographic) and summer convection. Mild in winter. Warm and humid in summer.
Tropical continental (Tc)	Originates over Sahara, North Africa. Tracks north and north west.	Brings hot, dry, stable heatwave conditions mainly in summer. Sometimes originates in Eastern Mediterranean and is more humid and prone to convectional storms. Heatwave often ended by violent storms – moisture from seas to south.

Table 1 The air masses that affect the UK

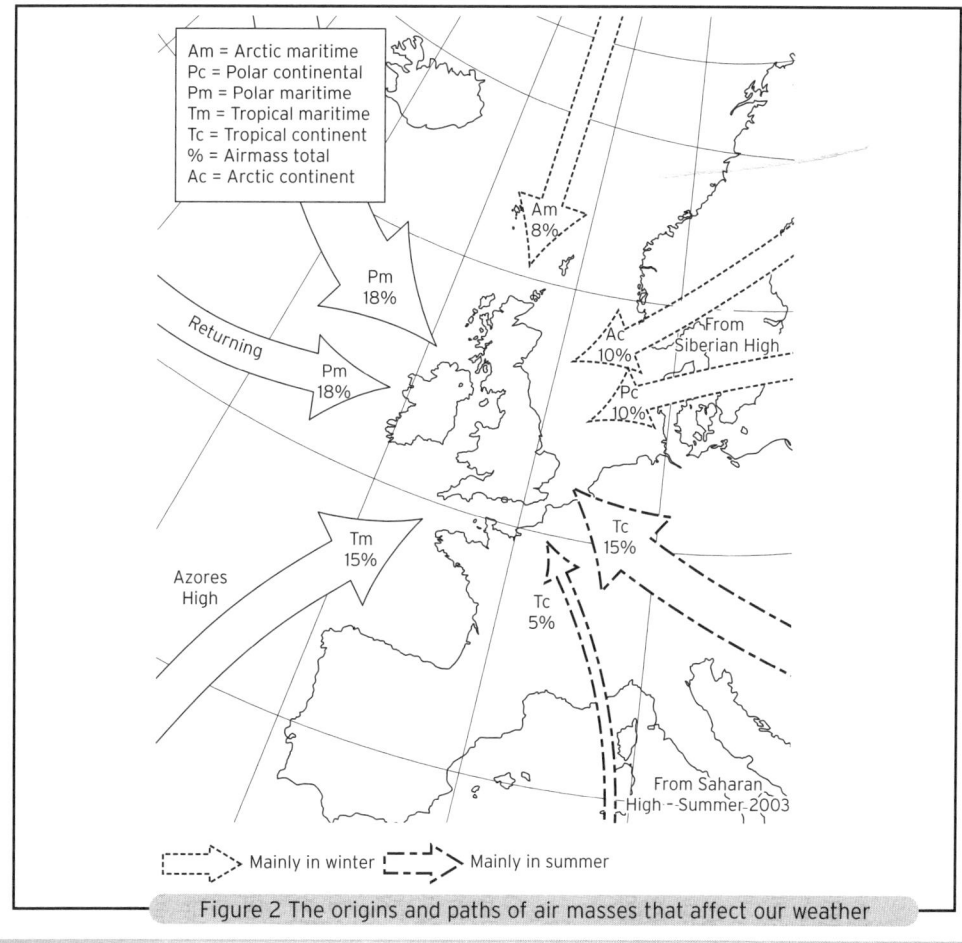

Figure 2 The origins and paths of air masses that affect our weather

The Polar Front Jet Stream

The Polar Front Jet Stream is a strong, narrow band of westerly winds in the upper atmosphere (12 000m) that effectively controls the weather in our latitudes. It is caused by heat contrasts between the tropical and polar air masses and is the mechanism for forming high and low pressure systems at the surface of the Earth. Consequently, it has a major influence on the Atlantic depressions that bring us our much needed year-round rainfall.

- Where the polar jet stream is moving south highs form, which is due to the intensifying of the high altitude jet stream. This leads to convergence and the descending air subsiding to give high pressure.
- Where the jet is moving north lows form, which is due to diverging high altitude air sucking up surface air. Therefore, at the surface frontogenesis takes place leading to the development of lows or depressions.

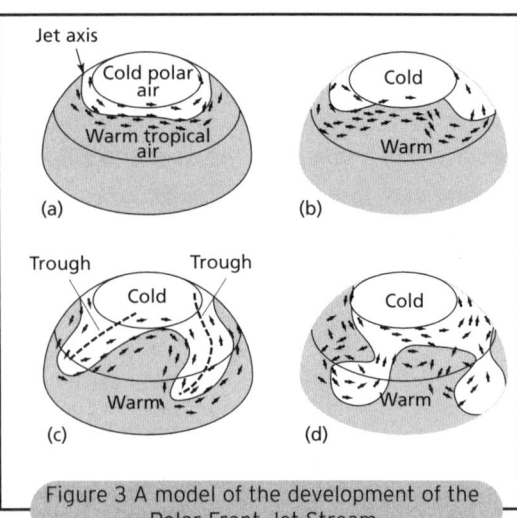

Figure 3 A model of the development of the Polar Front Jet Stream

The passage of a mid-latitude depression

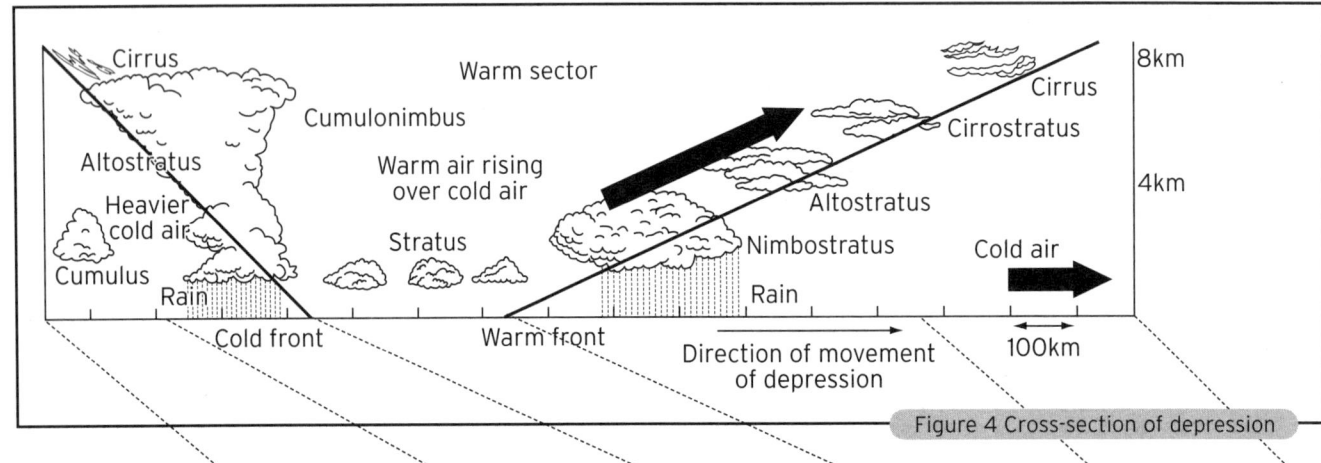

Figure 4 Cross-section of depression

	After cold front	Cold front	Warm sector	Warm front	Approaching warm front
Pressure	Rising	Sharp rise	Steady and low	Fall ceases	Steady fall
Winds	NW squally but decreasing Force 3-5	SSW veering NW Gusting to gale force 6-8	SW or S decreases Force 2-4	SSE veering SW. Strong. Force 4-5. Increasing strength.	SSE or SE Increasing in strength
Air mass	Pm or returning Pm	Changes	Tm	Changes	Pm
Temperatures	Cold 3-4°C in winter. Cool 13-15°C in summer	Sharp decrease at front of 4-5°C	Warm/mild 10-11°C in winter. Warm/hot 18-22°C in summer	Sudden rise of approx 5-6°C	Cool 6-8°C in winter and 15-17°C in summer.
Humidity and visibility	Falling rapidly. Good visibility except in showers	High until rains. Poor visibility but improving.	Relatively high, sultry in summer. Poor visibility.	High during precipitation. Decreasing visibility.	Rising slowly. Good visibility declining as cloud increases
Cloud cover and type	Diminishing, Fair weather cumulus	Towering cumulo-nimbus	Low, stratus, some clearing	Low, deep, nimbo-stratus	Succession from high cirrus – cirro-stratus, to low stratus
Precipitation	Heavy showers	Heavy rain, hail (snow and sleet), thunder	Clearing drizzle	Steady continuous rainfall. Maybe snow in winter turning to rain.	No precipitation until front close

Table 2 Summary of the changes that take place in the weather during the passage of a depression

1.2 Managing changeable weather globally and in the UK

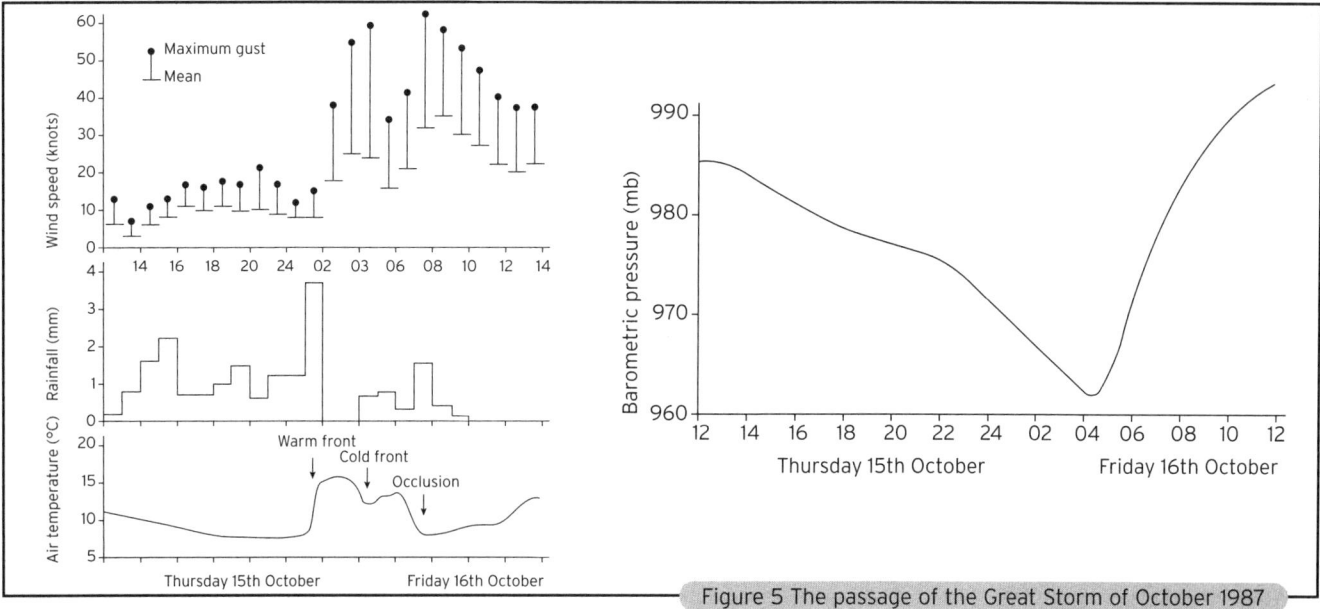

Figure 5 The passage of the Great Storm of October 1987

Low pressure systems

- Flooding
  (a) A continuous period of lows will result in a rising water table and overland flows, i.e. streams exceed bank-full capacity and flood (mainly between October and April). Many autumn floods in the UK are the consequence of such a period.
  (b) Cold front showers may produce flash flooding. Impact on crops, housing and industry on flood plains. Hail damage to crops in summer.
- Wind damage to buildings, trees and crops.
- Disruption to transport from fallen trees.
- Cost of claims to insurance industry. Increased premiums in subsequent years.

Reminder

Remember that an occluded front might be present on a weather map that you are given. What will the weather sequence be for an occluded front?

High pressure systems – anticyclones

	In Winter	In Summer
Air Mass	Pc	Mainly Tm – September 2003 but occasional heat waves of Tc – August 2003
Winds (always clockwise)	Light and from NE, E, SE depending on position of anticyclone.	Light from S, SE and SW depending on position of anticyclone.
Temperatures	Cold −5°C−+5°C	Warm to hot 18−25°C in Tm and hot to very hot 25−35°C in Tc.
Precipitation	Little. Snow and sleet on east coast.	Little except for thunder showers which signal 'break' of hot weather.
Humidity	Low except in areas near to exposed coasts.	Low but building up to very high in advance of storms.
Clouds	Clear with some stratus giving snow and sleet. Fog may form in hollows and in autumn as well.	Clear with decreasing visibility as humidity rises. Poor visibility in intense storms.
Timescale	Several days to 2−3 weeks.	Several days up to a month or more.
Impacts	Fog and smog on transport; Ice on transport − black ice, ice on lines; Increased power consumption; Access problems for those reliant on public transport; Health problems from smog; Blocking causes lows to move south of UK bringing rain to Southern Europe (winter rain of Mediterranean climate).	Flash floods; Photochemical smog leading to health problems; Pollen-related illnesses increase; Drought due to 'blocking effects'; Increased power consumption for air conditioning; Summer goods and foods increased sales; Blocking diverts lows to north − rain in Northern Scotland while drought in England.

Table 3 Types of weather associated with anticyclones, such as fog, frost and drought. However, they can vary from this model format especially in Spring and Autumn

 Can you draw the evolution of a polar front jet stream path and the creation of Rossby Waves? If possible try to memorise a three-dimensional version of the diagram because it will show the relationship between the waves and low (cyclonic) and high (anticyclonic) systems.

Reminder

You will need real examples of all the various impacts of low and high pressure on people, the economy and the environment.

1.3 Seasonal variations in climate

Key questions for this section:
- Why do seasonal variations in climate occur?
- What management problems does the seasonality of climates cause?

You will be revising:
- Seasonal variations in climate
- Climate types
- Their effect on people
- El Niño /La Niña.

This unit is about longer term or seasonal changes and their consequences for people. It is about seasonality, not just in the tropics although that is the most common place to study its impact. The monsoon is another form of seasonality which you might cover.

Seasonal variations in climate – seasonality

Your understanding of seasonal variations in climate should focus on the movement of the heat equator, as well as the ITCZ that affects the weather of both the tropical rainforest and savanna biomes (see 1.5, pp 24–27).

The classic example of the impact of seasonality on weather patterns and vegetation is that from the Tropic of Capricorn to the Equator in North Africa.

Climatic type	Air mass Summer/ Winter	Climate characteristics	Example	Ecocline	Approx. Latitude
Desert	Tc	30–40°C, Range 24°, Precipitation < 40mm, drought, high pressure Hadley cell	Sahara, S Algeria, N Niger and N Mali	Thorn scrub, some cacti. Oases support more grass.	16–20°+
Open Savanna	Tc/E Tc	26–30°C, Range 15° Precipitation 250–600mm, 7 months no rain.	Sahel, Burkina Faso	Tall tussock grasses and drought resistant trees	10–16°
Closed Savanna	E Tc	23–26°C, Range 12°, Precipitation 900–1500mm, 3–6 months Precipitation exceeds potential evapotranspiration	N Ghana, N Nigeria (Kano)	Woody shrubs + grass, drought and fire resistant, NPP Primary Productivity 0.9kg/m²/yr. Biomass 4kg/m²	5–10°
Equatorial	E	25–31°C, Range 6°, Precipitation >2100mm, 6–9 months Precipitation exceeds potential evapotranspiration.	West African Rainforest	Broadleaf evergreen, continuous growth, layered to 40+m high, diverse species, NPP 2.2kg/m²/yr. Biomass 45kg/m²	0–5°

Table 1 Climatic characteristics in North Africa from the Tropic of Capricorn to the Equator

Reminder

InterTropical Convergence Zone (ITCZ)
This is an area of slack winds and low pressure and high precipitation around the Equator where the Trade Winds from the NE and SE converge. It is approximately in the area of maximum surface temperature, the **heat equator**, and instability. ITCZ migrates particularly over land rather than over the oceans. This migration gives rise to **seasonality** – a feature of savanna climates and the Monsoon.

Monsoon climates
Areas where the winds show a marked seasonal change in direction and where the onshore winds bring a season of heavy rains. The pattern is caused by differential heating of land and sea and the poleward shift of the ITCZ.

1.3 Seasonal variations in climate

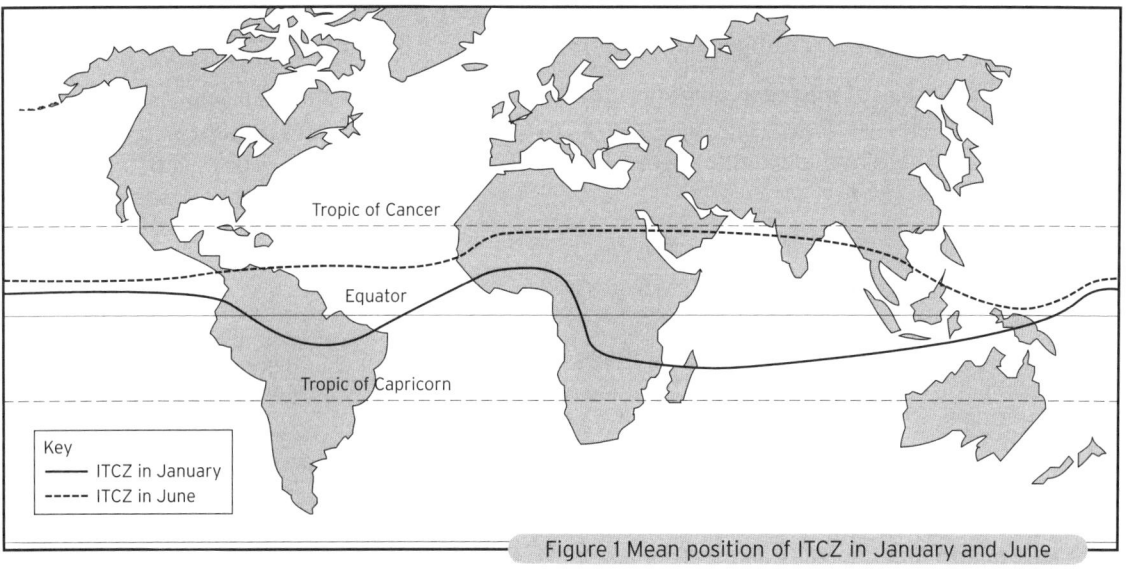

Figure 1 Mean position of ITCZ in January and June

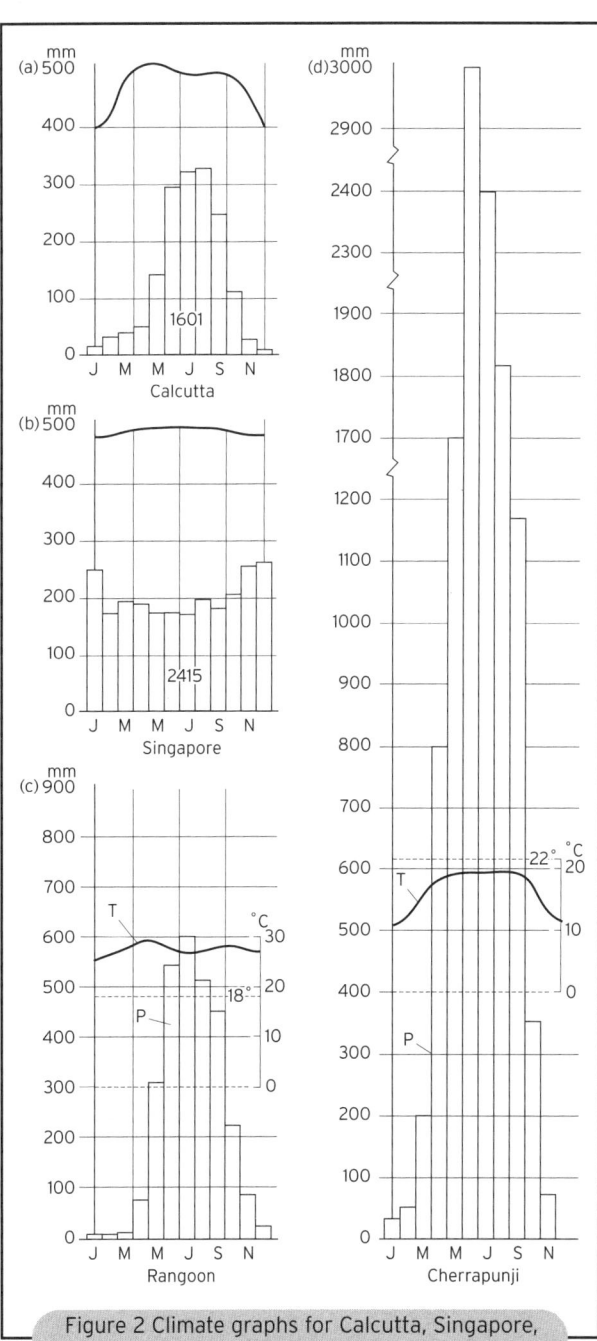

Figure 2 Climate graphs for Calcutta, Singapore, Rangoon and Cherrapunji

Djanet, Algeria	J	F	M	A	M	J	J	A	S	O	N	D	Tot
T (°C)	13	16	20	25	29	32	31	31	29	24	18	14	
P (mm)	0	3	3	0	0	3	0	0	3	3	0	3	18

Tahoua, Niger	J	F	M	A	M	J	J	A	S	O	N	D	Tot
T (°C)	23	26	30	33	34	32	30	29	30	31	27	24	
P (mm)	0	0	0	5	15	56	99	130	56	8	0	0	369

Kano, Nigeria	J	F	M	A	M	J	J	A	S	O	N	D	Tot
T (°C)	23	24	27	28	27	26	25	23	24	25	24	23	
P (mm)	0	2	13	65	150	180	215	300	270	75	2	0	1272

Table 2 Climate data for three stations in N. Africa

Reminder

You should be able to tell which climates are represented by the three sets of climate data in the table.

Reminder

You should be able to give an explanation for each of the climate graphs.

Seasonality occurs south of the Equator especially in Zambia, Malawi, Zimbabwe and Mozambique in Africa, and Brazil, Paraguay and Uruguay in South America.

Seasonality may also be studied in areas of **monsoon climates**. The graphs in Figure 2 illustrate the characteristics of Singapore, Rangoon, Calcutta and Cherrapunji and the transition from an equatorial climate to the extreme case of the monsoon where relief has exaggerated rainfall totals.

> **Reminder**
>
> There is more about **seasonality** on pp 36–40 of Bob Digby's (ed), *Global Challenges*, Heinemann 2001.

Issues of seasonality

1 Seasonal drought in Tropics

Examples include Ethiopia 1980s; The Sahel and Mozambique 1990s; and Zimbabwe 2000s.

Causes

- The ITCZ does not progress as far north or south drawing in wet equatorial air.
- The effects of the El Niño cycle (see below)
- Political boundaries stop migration of nomads and make management of international river water complex.
- Increased levels of herding and animal populations removing vegetation.
- Demands of cash cropping using irrigation.
- Removal of trees and scrub for firewood leaving soil exposed to sun.

> **Reminder**
>
> It is possible to discuss some aspects of seasonality with reference to the climate of the UK.

Solutions

Seasonality and variability of rainfall cannot be controlled. Possible alternatives include:

- Political solutions, e.g.: water agreements; encouraging farmers to pursue sustainable agriculture rather than irrigation-based cash cropping
- Improved wells (Water Aid do target this type of improvement) and water retention on land
- Improved animal husbandry and cultivation of biofuels for energy
- Managing waste water – costly
- Aid, such as BandAid and disaster relief for extreme events such as Mozambique flooding.

2 Drought in Europe 2003

Causes

- Air masses, such as Tc, dominated and remained further north than normal blocking the passage of depressions
- Tc is very dry air having passed mainly over land
- Tm air is also very slow moving. It lost humid characteristics when relatively stationary over NW Europe.

El Niño /La Niña cycle or El Niño Southern Oscillation (ENSO)

This is a challenging phenomenon whose effects are to alter the seasonal weather patterns over large parts of the world, not just the area where the variations in ocean currents and weather are occurring.

El Niño was first used to describe the appearance of a warm current off the coast of Peru that replaced the cold Humboldt current. Today, it refers to the large-scale changes in ocean currents and weather that occur across the Pacific Ocean between Indonesia and Australia and Latin America. A complex cycle involving **La Niña,** a cold event that often follows, is given the term **El Niño Southern Oscillation (ENSO)**.

El Niño /La Niña cycles were first recorded in 1532 and more recently in 1982–83, 1986–87 and 1997–98 (see Figure 3). In the past 100 years there have been 23 occurrences.

1.3 Seasonal variations in climate

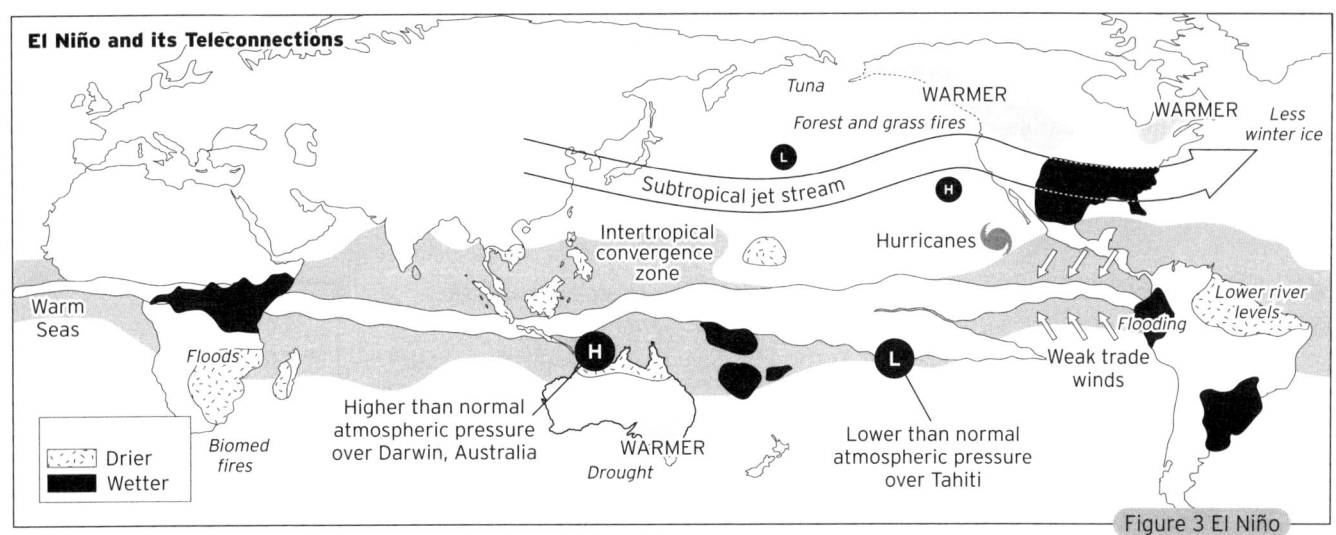

Figure 3 El Niño

El Niño (the Christ child)

Normally, the Trade Winds push water westwards (60 cm higher in the West than in the East Pacific). The water that is pushed westwards heats up so that water temperatures in the West Pacific are high. There is a lot of evaporation which leads to high rainfall totals in Indonesia and the Philippines. The movement of water west permits colder, nutrient-rich water to rise off the Peruvian coast and the coolest tropical waters are here.

During an El Niño year, there is a large-scale weakening of the NE and SE Trade Winds over the Pacific with very low pressure over Tahiti and very high over Australia. There is less warming of the waters of the East Pacific. The **sub-tropical jet stream** is further south and takes storms eastwards with the result that flash flooding is common.

> **Reminder**
>
> Be prepared to draw a similar map for La Niña

Characteristics
- It may be related to **global warming** but the evidence is not proven
- It occurs every 2–7 years with an average of every 4 years
- It lasts 12–18 months
- There is a southern oscillation, i.e. swings in sea level atmospheric pressure.

Environmental consequences
The trigger mechanism is not clear. Some of the consequences of El Niño years have been:
- The Pacific pressure gradient reverses giving rise to a complex reversal of the warm current towards South America

Human consequences
Some examples include:
- Loss of anchovy fisheries in Peru
- Loss of crops and livestock from flooding and fires

Teleconnections have been placed on figure 3

La Niña ENSO cold event

Characteristics
- Normally follows El Niño (15 times in past 100 years)
- Pressure abnormally low in West Pacific and abnormally high in East Pacific
- Sea temperatures are much lower in East Pacific
- Dramatic rise in sea temperatures in West Pacific
- Trade winds are stronger than normal
- Cold tongue extends up to 5000km from Ecuador to Samoa.

>  Write your own account of the causes and effects of the summer 2003 drought in West Europe. Try to introduce causes and effects on people, the tourist and manufacturing economies, agriculture, and transport. Why were people able to cope?

1.4 Climate change

Key questions for this section:
- What is climate change?
- What are the implications for environments and people?

You will be revising:
- Short term climatic changes
- The greenhouse effect and global warming
- Potential solutions at various scales
- Ozone.

Geographers often study the impact of change whether it be short or long term. Some of the most critical debates about the future of our planet surround our impact on short-term climate change. This section tests your knowledge of some of those crucial challenges.

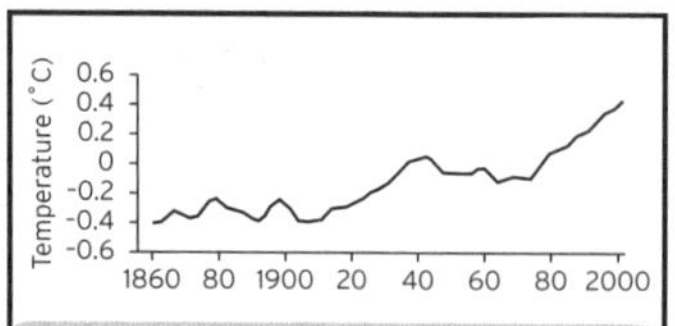

Climate change

Climate change can be **long term**, such as the changes that have occurred since the last ice age and events such as the little ice age in the sixteenth century when the Thames froze over. Colder winters lasted through until the mid-nineteenth century in the UK.

Short-term climate change normally refers to the changes that have occurred in recent history where there are accurate measurements of the changes. It includes our ability to extrapolate trends forward into the future. It is often discussed under alternative terms – **Global Warming** and the **Greenhouse Effect**. The ten warmest years on record have occurred since 1987. Global surface temperatures have risen 0.6°C since 1900 (Figure 1). Forecasts for 2100 in 1995 predicted 1.0°–3.5 °C; by 2001 this had risen to a predicted +1.4 °–5.8°C. Change is faster in the northern hemisphere because it contains more land, which heats up faster than the Southern hemisphere where the expanse of ocean water changes temperature more slowly.

The Enhanced Greenhouse Effect

Scientists believe that people have increased naturally occurring gases through industrialisation, urbanisation and the increasing intensity of agriculture. These gases stop long-wave radiation returning to space and so the planet heats up. (See Ozone on pp. 23)

What are the greenhouse gases?

They comprise CO_2 (55 per cent), methane (15 per cent), nitrous oxide (6 per cent), hydrofluorocarbons and perfluorocarbons (17 per cent) and sulphur hexafluoride (7 per cent). They are measured in CO_2 equivalents.

Figure 1 Global temperatures[1]

[1] Combined annual landsurfaces, air and sea - surface temperatures from 1860-2002 relative to 1961-1990

Reminder

Climate change
This has been the norm for millions of years. Scientific knowledge in the recent past is able to show that climates are variable. The rises and falls in average temperatures might be due to **Milankovitch cycles** produced by variations in the Earth's orbit, the Earth's tilt and the Earth wobbling on its axis

The Greenhouse Effect
A process that retains some of the heat escaping from the Earth and keeps the planet warm enough for life.
The Enhanced Greenhouse Effect is the increased impact of human activities on the process.

Global warming
The consequence, especially, of the Enhanced Greenhouse Effect.

Reminder

Remember to refer to Bob Digby's *Global Challenge*, Heinemann 2001 if you need to fill in some of the details.

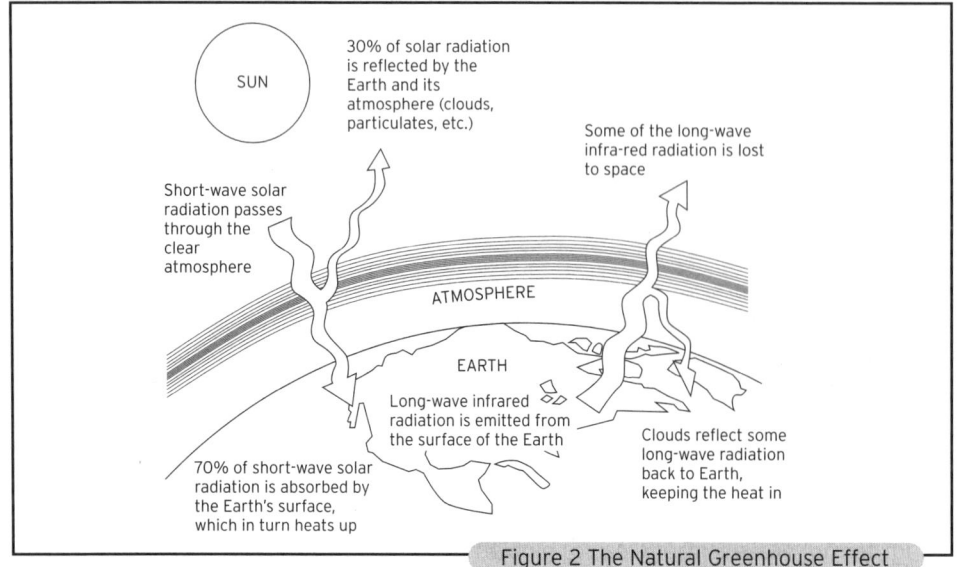

Figure 2 The Natural Greenhouse Effect

1.4 Climate change

Who produces the gases?
The developed world. The average US citizen's CO_2 output is 20 times that of an Indian and 300 times that of a Mozambican. The latest forecast is that newly industrialising Asia-Pacific will replace the OECD states (Organisation for Economic Cooperation and Development which has approximately 30 MEDC members) as the main source by 2015.

Alternative explanations for global warming
Scientists claim that at least 20 years more study is needed before it is certain that the Earth is warming. Sun spot activity may account for variations and the current changes in climate. The Global Climate Coalition set up by large industries in 2000 casts doubt on the theory and science of global warming.

Winners and losers from climate change in the UK
1. Pulp, paper and forestry industries benefit from increased yields and potential revenues from carbon sequestration or biomass energy.
2. Transport. Public transport may benefit from increased support. Aviation will face increasing costs.
3. Leisure. Warmer temperatures in northern Europe although winter sports may suffer.
4. Insurance. Heavy risks from impacts of climate change such as storms and flooding.
5. Oil and gas companies. The core product is implicated in climate change and therefore demand may be suppressed. Climate-related boycotts may increase as has happened against Exxon and BP.
6. Water. Companies may need to spend more to manage floods and droughts. They may also need to store more water and use it more efficiently.
7. Pension funds may be threatened by knock-on effects, e.g. investments in energy and insurance not performing well.
8. Drier, hotter summers (1995 and 2003) and wetter winters especially in the south east with increased seasonality. More extreme events such as flash floods in summer. Possibility of stormier weather.
9. Need to heat buildings will decrease but need to cool buildings will increase.
10. Soil moisture levels may decrease in summer necessitating more irrigation.
11. Pressures on, and delays to, emergency services may increase due to extreme events though warmer winters may place less demand on hospitals.
12. Land losses where coastal flooding is frequent. Schemes to permit reversion of land to marsh and abandoning land to the sea are being proposed for many low-lying coasts, e.g. Selsey area of West Sussex.
13. There is a suggestion that warming may have the opposite effect. Melting ice will introduce fresh water into the oceans, which will affect the pattern of ocean currents and may stop the flow of the Gulf Stream. Should this occur, the Meteorological Office models predict that NW Europe will be much colder with sea ice in winter.

Global effects of warming

- Models of change suggest 9–68cm rise in sea levels by 2080 (see Figure 3). Rise 1900–2000 was 10–20cm.
- One third of Bangladesh would be flooded by the sea and at least 20 million made homeless.
- Tuvalu and Kiribati in the Pacific would disappear – a member of the Alliance of Small States who are trying to pressurise the MEDCs to do more to combat global warming.
- The Gulf Stream will weaken due to input of more fresh water from glaciers lessening water density but it will not have the effect of cooling the UK for over 100 years.
- Collapse of Larson B ice shelf in 2002 released an iceberg the size of Jamaica into the Southern Ocean. The scale of ice loss, if continued, would contribute to the increase of fresh water in to the oceans (see figure 3).

Potential solutions

Global solutions

International agreements
You should know what the various international agreements have and have not achieved:

Reminder

Think about why small states are worried. What power do they have or not have and why can they not do more?

Global challenge

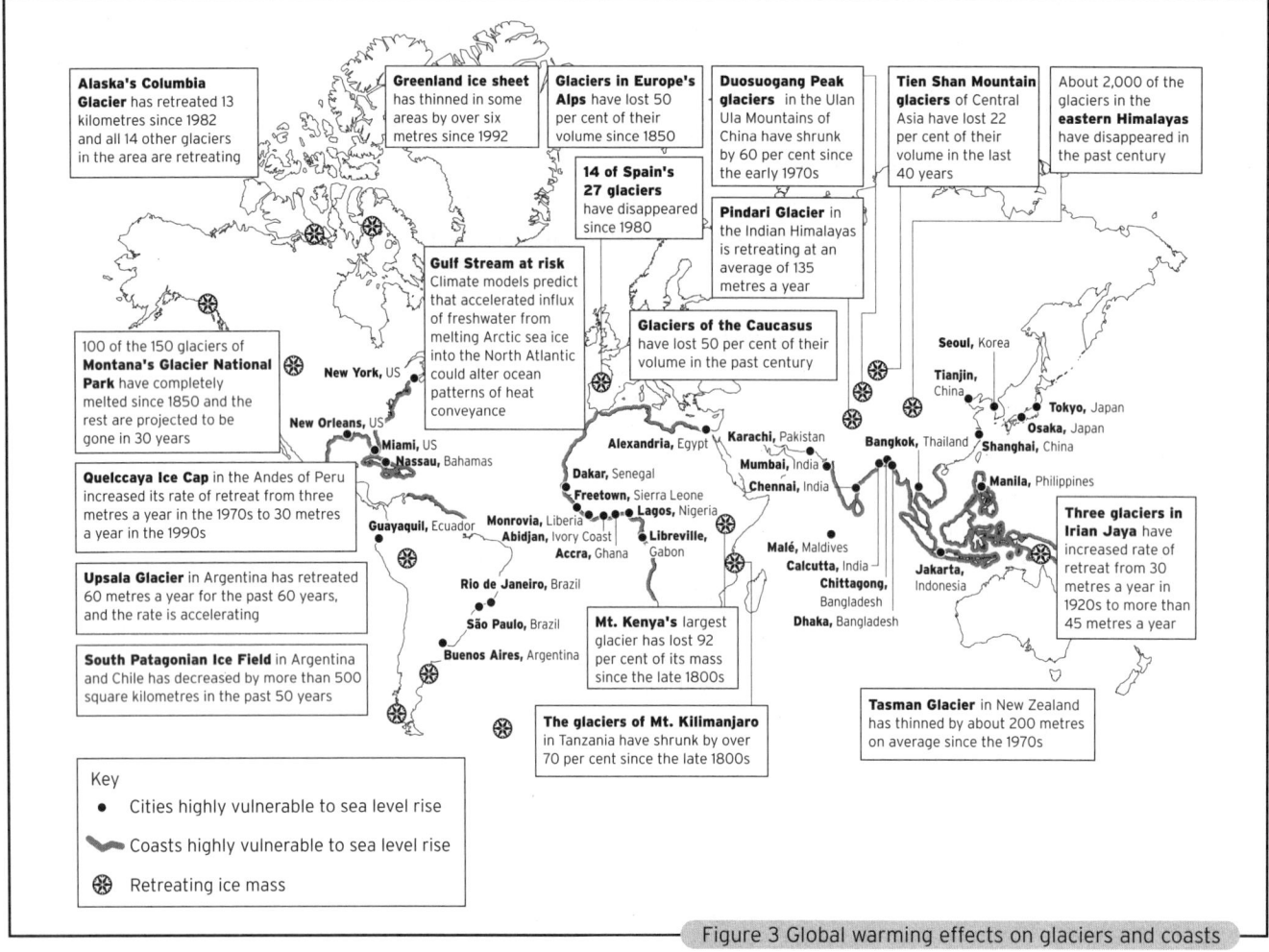

Figure 3 Global warming effects on glaciers and coasts

- **The Intergovernmental Panel on Climate Change (IPCC) 1988** The main body based in Switzerland that monitors, researches and proposes policies regarding climate change.
- **Montreal Protocol 1989** Agreement on substances that deplete the ozone layer. In 2003 the USA refused to stop use of methyl bromide, a potent insecticide and ozone-depleting chemical, thus challenging this treaty. Success to date because MEDCs were assisting LEDCs to meet cost of switching to non-ozone depleting chemicals.
- **Rio Convention on Climate Change 1992** Set first targets. 187 countries signed. Agenda 21 principles established. Six greenhouse gases including CO_2 emissions were no greater in 2000 than 1990. The USA and Australia did not sign the convention.
- **Kyoto 1997** 100 countries signed up – emissions to be reduced to at least 5 per cent below 1990 levels by 2008–2112. 60 per cent reduction to achieve standstill. Introduced **carbon credits** which enable plants to buy emission rights from less efficient plants in less developed countries or pay for an area of forest to be retained for 20 years. Allows the polluters to continue polluting. EU endorsed at Marrakesh 2001 target – 8 per cent. Kyoto not ratified by US (2001) because Bush's 2000 election funds came from many US oil multinationals such as Exxon who opposed it. US accounts for 36 per cent of emissions.
- **The Hague 2000** Talks where USA, Australia and Japan wanted bigger carbon sinks. Opposed by EU.
- **Johannesburg Earth Summit on Sustainable Development 2002** mainly focused on world poverty

Carbon sinks

- Forests and farmland that naturally soak up CO_2 and therefore lower greenhouse gas levels. Vegetation and soils absorb 40 per cent of emissions. They are not a long-term substitute for cutting CO_2 emissions because emissions are rising faster than the ability of soil and vegetation to cope.

Regional solutions

- Retain threatened forest, e.g. Meso-American Biological Corridor that extends south from Mexico to Panama, by protecting it.
- Injecting water into old oilfields in those regions with 'exhausted fields' to aid recovery of remaining oil. It is dangerous as they can leak and kill, e.g. California.

National solutions

The British target is 23 per cent reduction in CO^2 emissions below 1990 levels and 10 per cent of electricity from renewable sources by 2010. Possible ways of doing this are as follows:

- Cut private vehicle use, which has only been done once on a large scale and that was during World War II.
- Improve generation and use of energy – the best way to cut emissions. The biggest culprit is not power but cement works – an industry that is basic to all construction.
- The Climate Change Levy uses taxation to encourage a cut in the output of gases through increased fuel efficiency. Yet government is spending more on roads and subsidises drilling for oil – is this inconsistent?

The US solution is to increase development so that people can afford less polluting technologies. This solution is subscribed to by major oil companies.

> **Reminder**
>
> Think about why oil producers take this stance.

Local solutions

More recycling; greater energy efficiencies in home, e.g. lower washing temperatures, more efficient heating,

Ozone

Ozone occurs a) at ground level near surface, and b) as upper atmosphere ozone in the troposphere 20–50km up in a band 25km thick where it is known as the **ozone layer**.

Holes in the ozone layer

Upper layer ozone is a relatively small proportion of the atmosphere and can be damaged easily. It shields us from the sun's ultra-violet radiation which can increase surface temperatures. Holes were noted over Antarctica in the 1970s and more recently over the Arctic. Holes form where the ozone has been destroyed by volcanic dust and the subsequent increase of carbon monoxide. People have damaged the layer through the use of CFCs, in for example refrigerators and plastics, that have a life of over 100 years and destroy ozone. The effects of a damaged ozone layer are increasing levels of eye damage and incidence of malignant melanoma. It has effects on ecosystems at the lower stages of the food chain because ultraviolet radiation affects germination.

> **Reminder**
>
> Be careful not to confuse the effects of the two types of ozone and in particular the role of the ozone layer, i.e. the tropospheric ozone, in global warming. Ground level ozone is a hazard and a pollutant and best discussed as either a climatic hazard or a pollutant.

Build up of ozone layer

Ozone builds up in the troposphere due to emissions of hydrocarbons and nitrous oxides from power stations, vehicles and solvents. The outcome is that ozone levels are increasing over North America and Europe. This increase in the ozone layer helps to retain long-wave radiation and leads to global warming and is the enhanced greenhouse effect.

Ground level ozone

Forms mainly in summer in urban areas where the sources are concentrated and especially in natural basins such as Athens or Los Angeles. Anticyclonic conditions are the most conducive to its formation because the air may not move much. The ozone can react with sunlight to produce **photochemical smog,** which will affect health and damage rubber.

> **Web link**
>
> For more research into climate change and ozone, go to the following website and enter the express code 1552S:
> www.heinemann.co.uk/hotlinks

Quick Check: Do you know the effects of global warming on an LEDC?

1.5 Biomes, ecosystems and the threats to their survival

Key questions for this section:
- What factors are responsible for the global patterns of biomes?
- What is the importance of global ecosystems?

You will be revising:
- What defines a biome and an ecosystem
- The unique characteristics of forest, grassland and marine biomes
- Ecological footprints
- The threats posed to ecosystems.

This is the second part of your studies for Section A 'The natural environment' in the examination. It is important because it has relevance if you study 'Wilderness environments' for 6475/1 (see 2.4, pages 109–118). It also has a major input into the cross-unit questions (see page 74) and needs careful study. There is the option within the unit to study forests or grasslands or marine ecosystems. It probably pays to study one in depth and to have some knowledge of a second ecosystem type.

> **Reminder**
>
> What you select to study here may link to your options (see 1.6–1.8, pp 28–34)
>
> You will need to have studied one of the biomes in Tables 1–3. Your college probably made the selection for you but you may have the chance to study an alternative biome.

Biomes and ecosystems

Every **biome** is defined by unique flora and fauna within a specific climatic region. Other than climate, the flora and fauna of a biome are affected by factors such as soils, precipitation moisture levels and sources of food. There are ten major land-based biomes (Figure 1).

An **ecosystem** is composed of the often complicated links between the plants and animals, and their environment (climate, soils). It can be studied at all scales from the biome to a small area within a biome, e.g. an oak wood within the temperate forest biome. Ecosystems may become **stable** when all the components are in equilibrium or are changing extremely slowly. Ecosystems can become **unstable** when there is a major disturbance such as a drought.

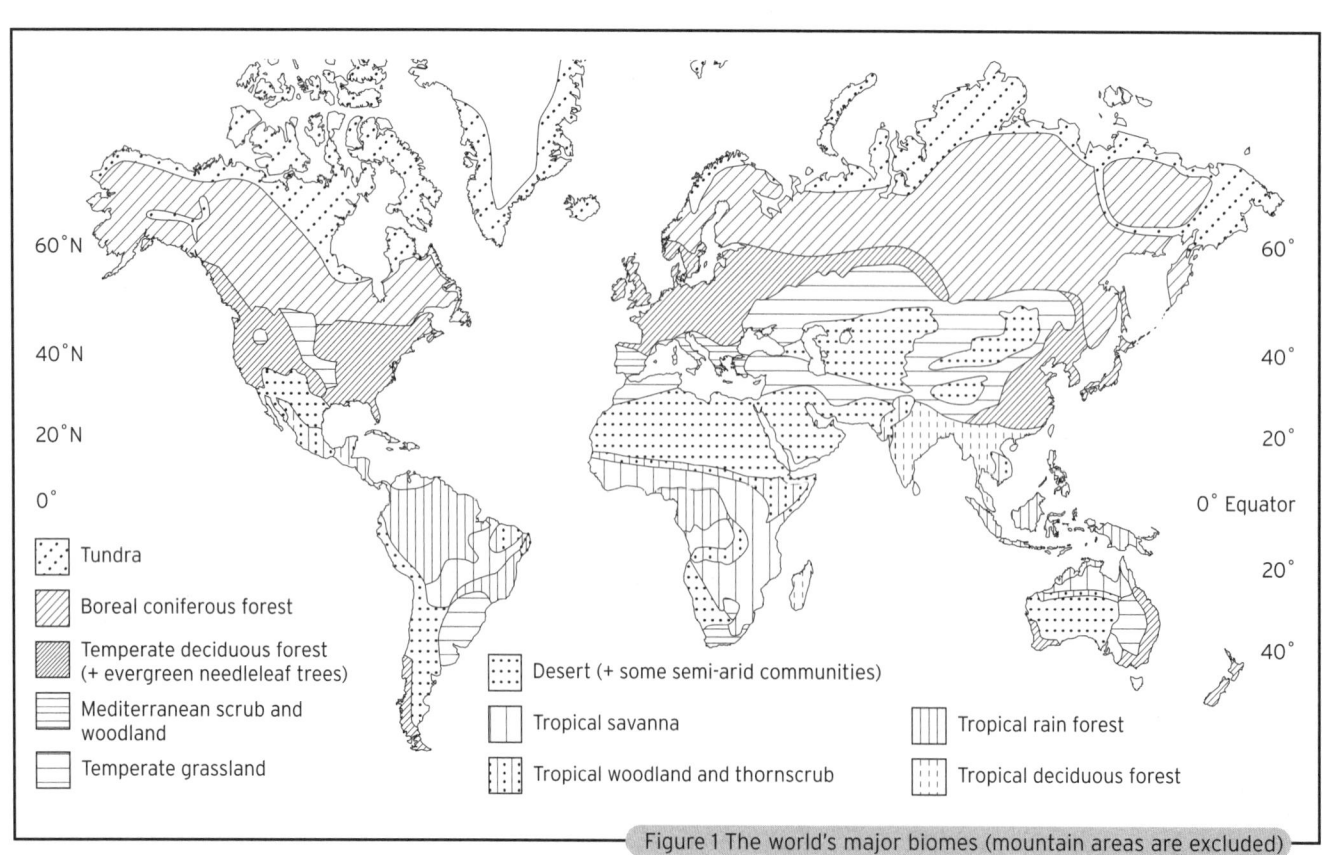

Figure 1 The world's major biomes (mountain areas are excluded)

1.5 Biomes, ecosystems and the threats to their survival

The table below gives the main characteristics of three forest biomes, two global grassland biomes and major marine ecosystems

	Tropical	Temperate	Boreal or Taiga
Climate	Equatorial at the ITCZ	Temperate Western Margin	Cold continental and sub polar. Short summer
Dominant air masses	Equatorial and some Tropical continental	Polar maritime and Tropical maritime	Polar continental, and Polar maritime with Arctic continental in winter.
Precipitation	Year round, 1200mm minimum up to 3000mm. High humidity.	Year round with winter max. 600–1800mm	Summer max., 550–650mm. Snow in winter – ground frozen.
Temps (average)/sunlight	27–33°C constantly hot. High levels of **evapotranspiration**	Range from 5–10°C in winter to 15–20°C in summer.	Range 0– -10°C in winter to 10–20°C in summer.
Characteristics	**Evergreen** with rapid growth. Species rich. Distinct layers. Buttress roots. Tall seeking light. **Epiphytes** (lianas) and **saprophytes** (fungi). Deciduous ecosystem variant with seasonality of rainfall.	Mainly **deciduous**. Stratified. Species rich.	Predominantly **coniferous**, slow growth. Little ground cover. Needle leaves reduce transpiration. Pollination and seed growth over 2 years
Nutrient cycling	Rapid – most energy stored in biomass	Moderate – soil and litter store significant	Slow.
Soils	Rapidly weathered, heavy leaching **laterites** or **red earths (ferrallitic** and **oxisols)**	**Brown earths** (cambi-sols) and podsols	**Podzols** with leaching 7 gleys
Species	Mahogany, ebony, rubber. Varies on each continent.	Oak, beech, elm, sycamore. Much fauna especially herbivores. Biodiverse.	Pines, firs and spruces
NPP or **Net Primary Productivity**	1500g/m²/year and more	650–2600g/m²/year	400–1900g/m²/year

Table 1 Forest systems

	Savanna	Temperate – Prairie, Pampas and Steppe
Climate	Tropical seasonal (wet/dry) (see 1.3 above)	Temperate continental – cold dry season and wet summer season
Dominant air masses	Tropical continental in dry season and equatorial in wet season. ITCZ movement critical in advent of wet season.	Polar continental in winter
Precipitation	Highly seasonal. Wet season may be very wet whereas drought very common in dry and may extend into wet season – 300–1600mm.	500–1200mm – evaporation high in summer.
Temperatures	23–28°C.	
Characteristics	Deep tap roots. Thick bark (against fire), waxy leaves. Wide range of fauna mainly migratory in search of water/ vegetation.	Matted root system resists erosion. Less dense in drier areas
Nutrient cycling	Most stored in soil.	Grasses decay rapidly in summer. Slow in winters due to cold.
Soils	**Ferralitic** but less leached than rainforests	**Chernozems (black earths)** – deep with upward movement of nutrients. Much organic matter.
Species	**Xerophytic** (drought loving) and **pyrophytic** (fire loving) grasses. Some trees, e.g. baobab and acacia palm (eucalyptus in Australia).	Tussock grass. Other grasses 55–65m high. Much altered by peoples e.g. native populations' grazing herds = **plagioclimax** vegetation. Big **herbivores,** e.g. buffalo.
NPP	900g/m²/year	600g/m²/year

Table 2 Grassland ecosystems

	Corals	Mangroves	Coastal wetlands
Where	Shallow tropical waters 30° N and S. 91% in Indian and Pacific oceans.	8% of world's coasts and 25% of tropical coasts.	S Florida Everglades, Camargue, Ebro Delta. Coastal tidal marshes such as Langstone Harbour, Suffolk estuaries in UK.
Water temperature	Warm 20–28°C; above 33° causes bleaching.	Warm as for corals.	Cool
Water state	Calm inside reefs. Less than 30m deep.	Calm but affected by tropical storms.	Calm but affected by tidal range.
Silt	Silt-free clear water. **Siltation** and pollution cause death.	Silt laden. **Symbiotic relationship** with coral – acts as silt filter.	Silt-laden salt marsh and mudflats.
Species	Wide diversity of coral polyps and algae. Invasive species a threat, e.g. starfish.	Very diverse low trees and shrubs. Salt tolerant. Spawning grounds.	Wading birds. Very diverse
NPP	2500–3000 g/m2/year	1600g/m^2/year	300–3000g/m^2/year

Table 3 Coastal ecosystems

Human impact on ecosystems – ecological footprints

Human impact is best measured by the **ecological footprint** showing the land area required to produce the food or, the wood that an inhabitant consumes in a year. Figure 3 shows the area of forest required to absorb the carbon dioxide produced per capita by burning fossil fuels and human activities. The greater the figure, the greater the ecological impact of the people in that country and the less ecologically sustainable their activities.

Reminder

If you have studied tropical rainforests for GCSE it might be better to select one of the other two forest systems to make sure you reach the correct standard of knowledge and understanding.

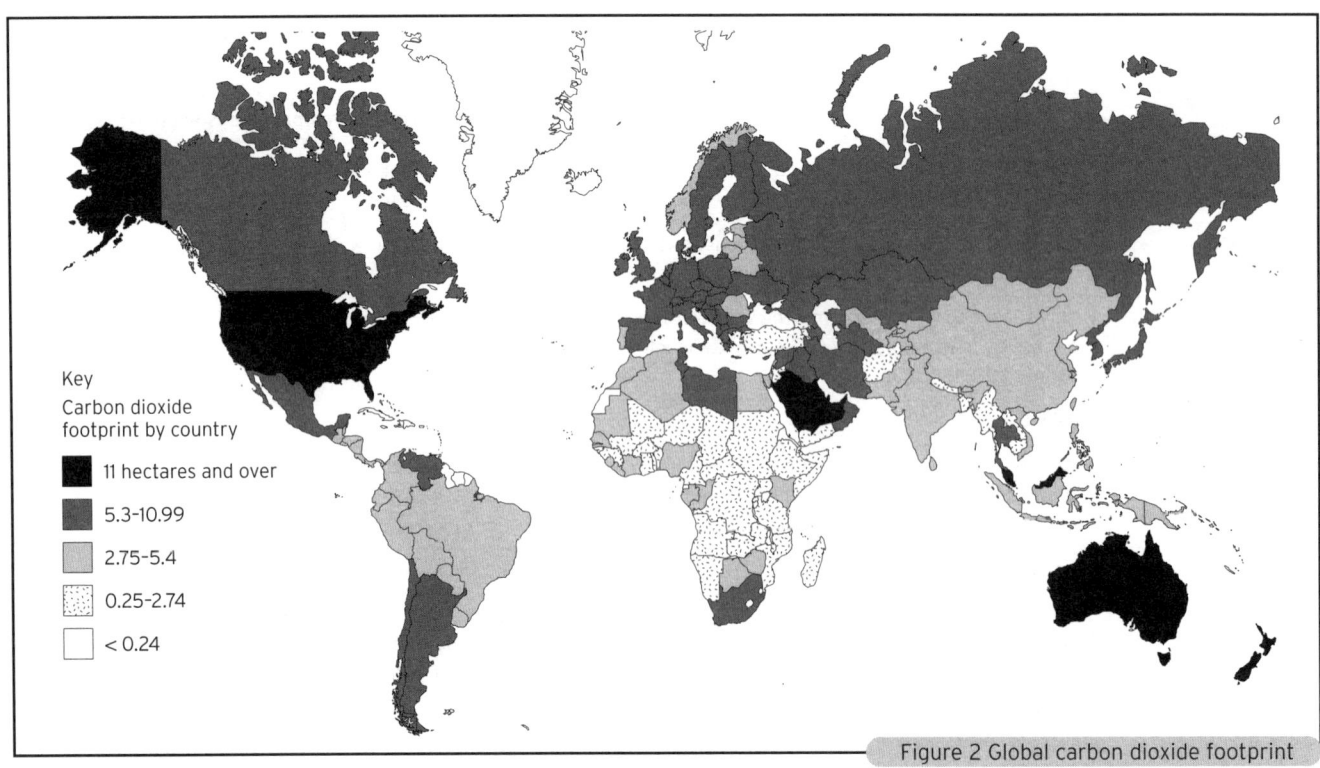

Figure 2 Global carbon dioxide footprint

Importance of ecosystems

You will need to relate these benefits to your chosen ecosystem. The list provides you with a checklist.

1.5 Biomes, ecosystems and the threats to their survival

Direct benefits

- Food – crops, grazing land, fish.
- Raw materials such as timber, bamboos, salt
- Genetic resources to improve yields, improve disease resistance
- Medicines
- Carbon storage especially in forests (global benefit)
- Jobs in agriculture and tourism
- Biodiversity as an insurance against climate change, drought and vegetation stresses
- Soil generation especially in grasslands
- Industrial use of timbers, grasses and of seaweeds (iodine).

Indirect benefits

- Erosion control especially by forests and grasslands
- Water purification and filtration
- Pollination
- Scenic enjoyment and spiritual significance
- Importance of mangroves to moderate storm impact on coastlines.
- Waste dilution and treatment.

> **Reminder**
>
> Make sure you have an example of the climate either as figures or as a graph for your selected ecosystem.

> **Reminder**
>
> Make sure you can add your own examples of direct and indirect benefits from your chosen ecosystems.

Threats to ecosystems

Threats	Forests			Grasslands		Marine		
	Tropical	Temperate	Boreal	Savannas	Temperate	Coral	Mangrove	Estuarine
Timber extraction								
Fuel-wood and charcoal								
Deforestation								
Fragmentation of landholdings								
Fire								
Biodiversity loss								
Desertification								
Overgrazing								
Cash cropping								
Overfishing								
Population growth								
Pollution								
Coastal development								
Tourism								
Recreation								
Mineral extraction								
HEP								
Port development								

Table 4 The threats to ecosystems from human activities. The shaded boxes are those where there is a definite threat to the ecosystem from human activities. In the case of marine systems the threats can be either on shore or off shore.

 Provide an example for all of the shaded boxes for each of the ecosystems on Table 4.

1.6 The degradation and conservation of the world's forests

Key questions for this section:

- Why has the degradation of the world's forests become such a global issue?
- Why are forest biomes so difficult to conserve?

You will be revising:

- Nutrient cycling and energy flows
- Threats to forest ecosystems globally
- Conservation issues for forests.

You are only expected to study ONE of 1.6, 1.7 and 1.8 for the examination. By far the heaviest content is in this section 1.6. If you are studying forests make sure that you are not relying on knowledge from GCSE or earlier. Remember that some of the knowledge that you have gained here may be useful if you are studying 'Wildernesses' (2.12, pp 81).

Chapters 6 and 7 in Bob Digby's Global Challenges will provide you with more detail. This section will remind you of various principles of ecosystems while focusing on different forest ecosystems. You will need to fill in the gaps for all three ecosystems

Characteristics of forests

The different flows, cycling and stores of energy within an ecosystem are key characteristics that aid our understanding of the threats to ecosystem stability.

Nutrient cycling

The diagram below (Figure 1) shows the cycling of nutrients in a boreal (coniferous) forest ecosystem.

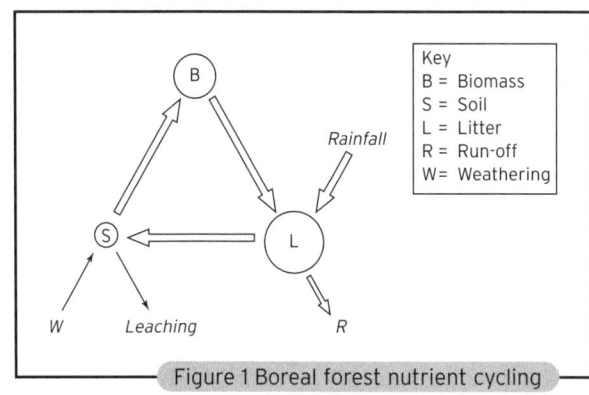

Figure 1 Boreal forest nutrient cycling

Energy flows

Energy flows through the **trophic levels** (Figure 2).

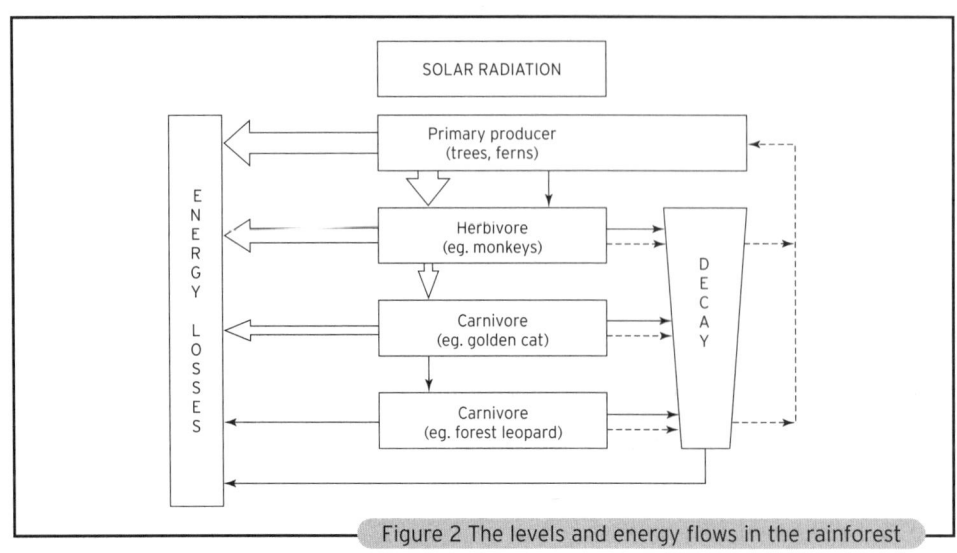

Figure 2 The levels and energy flows in the rainforest

> **Reminder**
>
> **Biodiversity**
> The range of flora and fauna species in a wide variety of **ecological niches**.
>
> **Plagioclimax**
> Where an ecosystem such as forest has been cleared and left to regenerate, e.g. after tribal slash and burn.
>
> **Trophic level**
> The level at which energy is transferred from one organism to another as a part of the **food chain**.
>
> **Climax vegetation**
> Attained when the vegetation is in harmony with the climate; will only change if there is a major change in the long term climate.

> **Reminder**
>
> Make sure you can draw the distribution of forest types on a world map (see page 81 in Global Challenges). The more accurate you are the more it will gain credit.

> **Reminder**
>
> Can you draw a similar cycle for the tropical rainforest?

1.6 The degradation and conservation of the world's forests

The structure of a tropical rainforest

Tropical rainforests are stratified. Layers include:

Floor The ground layer is dimly lit and often bare except for dead and decaying vegetation

Sub-canopy Trees and saplings with forest animals, limited undergrowth. **Lianas** link the layers; **epiphytes** are attached to trees.

Canopy Dense, continuous layer of tree crowns with most birds and insects

Emergent layer Trees which penetrate, or emerge from, the canopy.

This complex of layers, plants and animals forms the **climax vegetation**.

Forest degradation

The key theme of this section is **forest degradation** but you might have studied other forest types where the cases are different although the principles might be the same.

Tropical rainforests

The causes of rainforest degradation.

- **Slash and burn subsistence farming** – if farmers return too quickly to the same area it results in an increasing loss of nutrients despite burning which provides some nutrients. It is sustainable only if population of indigenous groups does not explode.
- Commercial activities such as **ranching** and **plantations,** e.g. oil palm in Malaysia, expose soil to nutrient loss over a longer period. Fertility declines and land abandoned.
- Commercial timber exploitation for hardwoods, especially prevalent in Indonesia and Brazil where 1 per cent of forest is being lost every year. Figure 3 shows deforestation in the Amazon.
- Erosion exposes soil to the elements leaving hard iron oxide layer which prevents vegetation growth.
- Opening up areas for settlement and communications, e.g. the Trans-Amazon Highway.
- Mining, e.g. Freeport Copper Mine, Timika, Papua

> **Reminder**
>
> Slash and Burn is subsistence farming and NOT the burning of forest to clear for plantation.

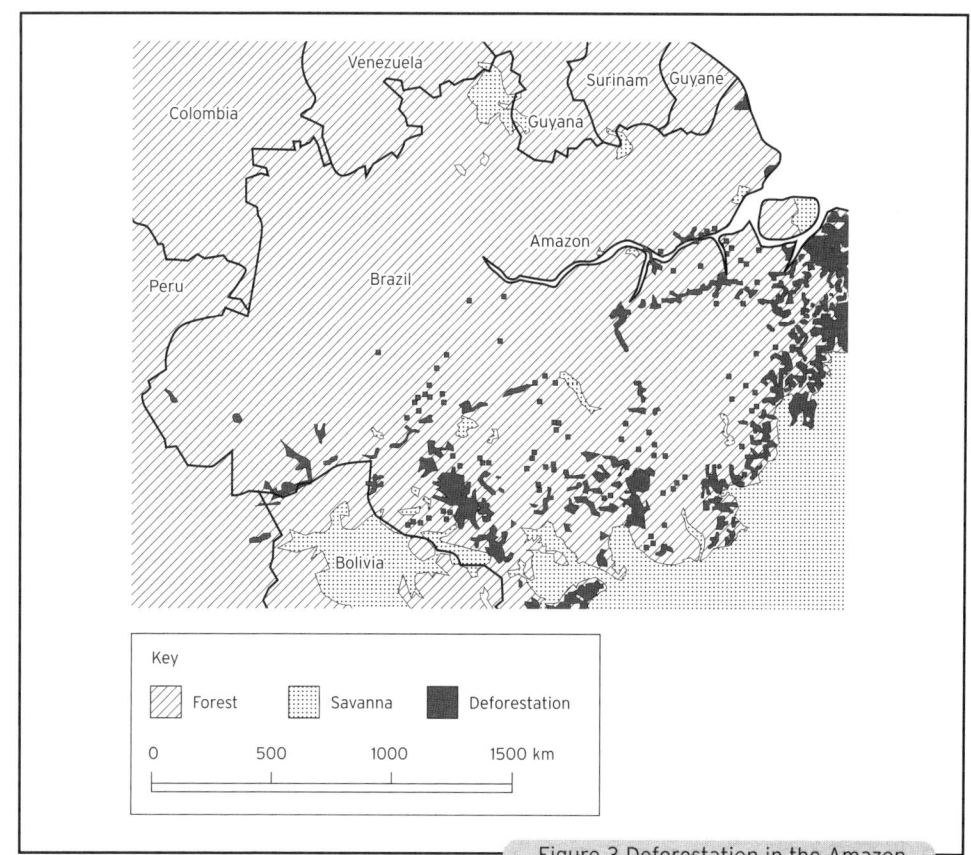

Figure 3 Deforestation in the Amazon

- Governmental attitudes to environmental conservation, e.g. the American Bush administration wishes to explore the Alaskan forests for oil.
- Threats to indigenous peoples:
  1. Diseases brought in by settlers
  2. Loss of lands as part of the slash and burn cycle
  3. Western culture.

Boreal forests

The degradation of most boreal forest in Northern Europe from, for example, logging, has been a threat for longer than to the tropical rainforest. But there are other more recent threats to contend with such as the pollution shown in Figure 4.

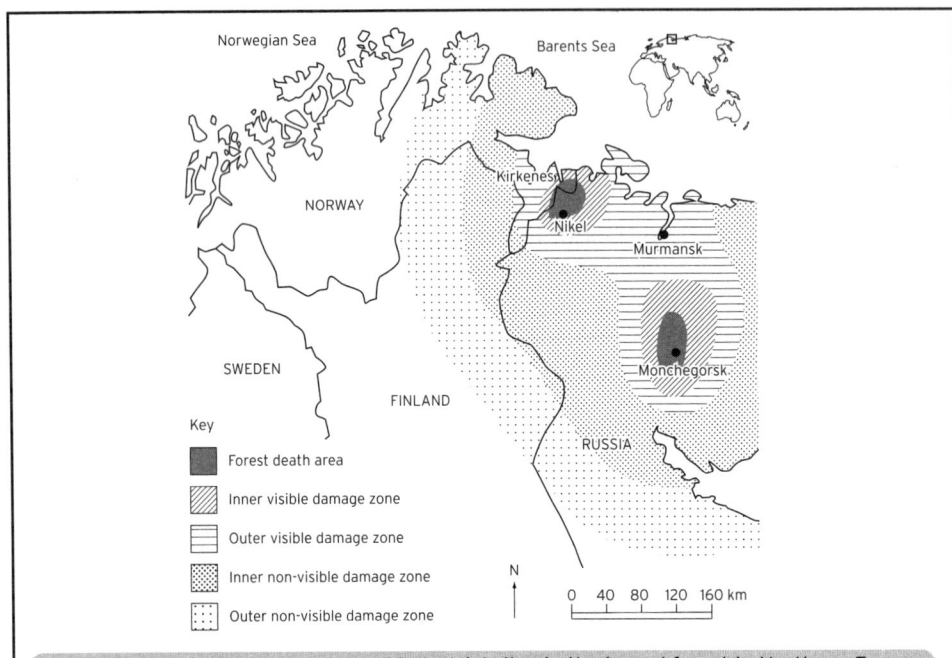

Figure 4 Air pollution damage round industrial sites in the boreal forest in Northern Europe

Solutions

The solutions here are for boreal (coniferous) forests but you could develop your own list of solutions for other forest types.

- Conservation – the law forbids destruction of young forest
- Sustainable management – especially in Finland where the products contribute 8 per cent of GNP. Logging in 2ha patches so that the total area of 350 000ha cleared each year is spread out and less noticeable. Thinning assists strong growth and represents half of all logging
- Forest conservation schemes – such as in forest parks and national parks, with footpaths and ski trails to maintain strong links between people and environment
- Agro-forestry – crops grown between rows of trees thus retaining some of the ecosystem while still permitting crop growth. Some of the trees may be plantation crops such as cocoa or oil palm
- Reforestation – takes between 60–120 years for trees to reach maturity (80 years for birch trees – slow growth due to climate)
- Encouraging recycling of timber products such as paper and board
- Governmental legislation – to underpin ecological sustainability, the preservation of biodiversity and environmental protection
- Greater control over pollution from timber processing plants.

However, further measures could reduce the competitiveness of forest products, e.g. in Finland.

1. Give the reasons for the difficulties conserving a forest biome.
2. Write notes suggesting the possible solutions to the threats to either tropical rainforests or deciduous forests.

1.7 The deterioration and desertification of the world's grasslands

Key questions for this section:

- What pressures lead to the deterioration and desertification of the world's grasslands?
- Why do grasslands prove difficult to manage?

You will be revising:

- The distribution, uses of and threats to savanna and temperate grasslands
- The causes of desertification
- Sustainable strategies for grassland conservation.

The second optional topic requires you to study two ecosystems: tropical savanna and temperate grasslands.

Characteristics of grasslands

Tropical savanna

See Table 2, 1.5 for some characteristics.

Trophic levels
1. Grassland and scrub
2. Herbivores – zebra, wildebeest, giraffe (selective grazing). Migrations away from dry grasslands in dry season. Numbers stable.
3. Carnivores – lion, hyena.

Nutrient cycling
Fire is used by peoples to clear areas and promote new growth by returning nutrients to the soil. The process can cause the dominant vegetation to change. Forty per cent of global biomass burning comes from savannas (see Figure 2).

Human uses
- Semi-nomadic grazing; overgrazing may occur because of population pressure, e.g. Fulani in northern Nigeria and Maasai in Kenya; movement is being restricted by government.
- Increasing cash crops such as cotton, tobacco; often the areas of European settlement in Africa.
- Tourism – game parks, e.g. Masaii Mara and Amboseli in Kenya; Kruger in South Africa.

Threats
- Erosion caused by overgrazing and trampling around waterholes.
- Unreliable rainfall due to non arrival of ITCZ.

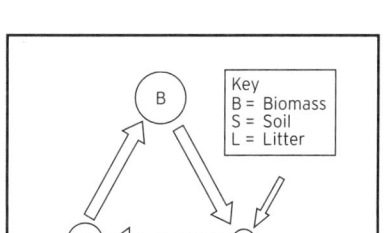

Figure 1 Savanna nutrient cycle

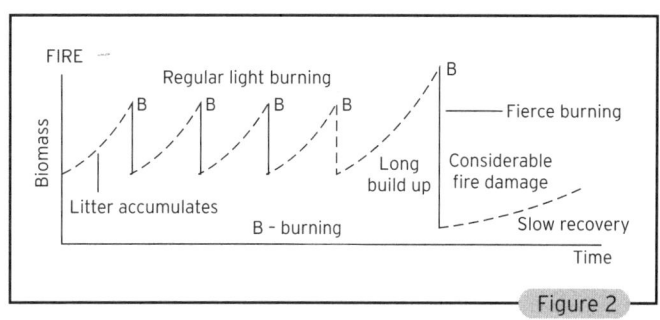
Figure 2

> **Reminder**
>
> **Desertification**
> The extension of desert landscapes particularly around the borders of deserts. It can be partly the product of climate change but is more often seen as man-induced processes that lead to soil nutrient depletion and a reduction of biological productivity.
>
> **Cash crop**
> Crops grown for cash – export earnings – rather than to feed the population.
>
> **Salinisation**
> The process of salt accumulation in the soil due either to evaporation of soil water or evaporation of irrigation water.

Temperate grasslands

These include the Prairies, North America; the Steppes, Eurasia; Pampas, Argentina; Veld, South Africa; East South Island, New Zealand. (See 1.5, page 24 for some characteristics.)

Trophic levels
1. Tussock grassland
2. Herbivores – buffalo, bison, caribou
3. Carnivores – wolves.

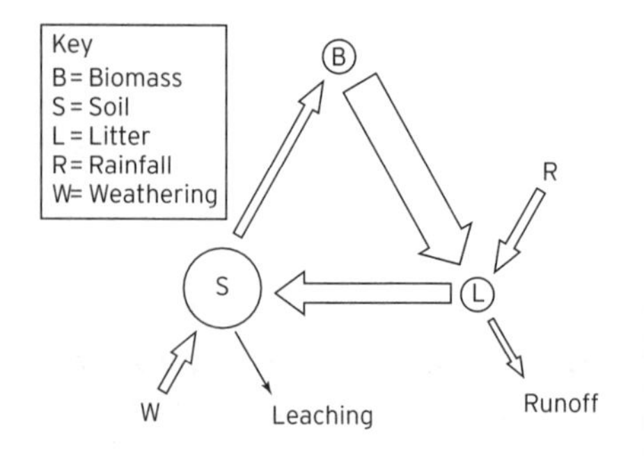

Nutrient cycling

Fire is used to promote growth and return nutrients to the soil. This has changed the soil's composition and structure. Fire can also be caused by controlled burning and lightning strikes. Arson may also occur.

Human uses

- Grazing of herds, e.g. prairies
- Modern cattle ranching, e.g. Pampas
- Extensive agriculture for cereals. These three uses have changed 100 per cent of North American prairies.
- Urbanisation
- Cowboy films
- Some tourism, e.g. Dinosaur Park, Alberta, Canada.

Figure 3 Temperate grassland nutrient cycle

Threats

- Soil erosion from wind, ploughed fields, e.g. the Dust Bowl in 1930s USA which was the subject matter for John Steinbeck's *The Grapes of Wrath*
- Gully erosion, e.g. US 'badlands'
- Over-use, over-harvesting and loss of fertility, e.g. Southern Ukraine
- Tornados and summer storms wreck crops
- Heavy snows in winter prevent grazing and make access difficult to feed herds
- Pollution from cities
- Lack of the continuous vegetation cover that enables species migration
- Species invasion, i.e. **non-native** species such as cane toads in Australia
- Rainfall variations and mismanagement
- Removal of scrub vegetation and trees in woodland savanna for firewood
- Schemes such as ground nuts which failed in Kenya because of inappropriate attention to the environment
- Loss of biodiversity due to agriculture simplifying the system, i.e. reducing species diversity to the single agricultural crop (Figure 4).

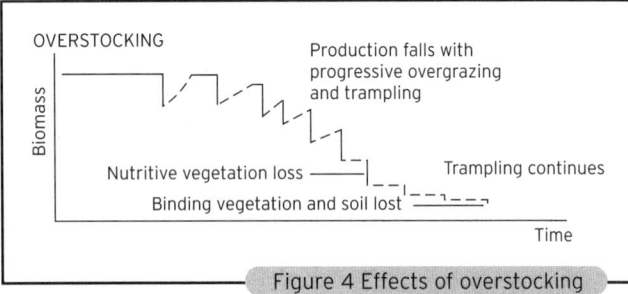

Figure 4 Effects of overstocking

- Diseases, e.g. sleeping sickness caused by the tsetse fly
- Population pressure
- Political threats, e.g. Zimbabwe
- Loss of biodiversity because some plants eaten out of existence leaving weeds as the dominant species.

Desertification of grasslands

This has caused the deterioration of most areas fringing dry lands and has resulted in the loss of usable land. Figure 5 shows the global pattern of deterioration. It tends to affect LEDCs and therefore most of the world's poorest people.

The total area affected by desertification is 3600 million hectares; an additional 6 million hectares are added every year. Grasslands store 33 per cent of the world's carbon. Destruction will reduce the world's capacity to store carbon.

1.7 The deterioration and desertification of the world's grasslands

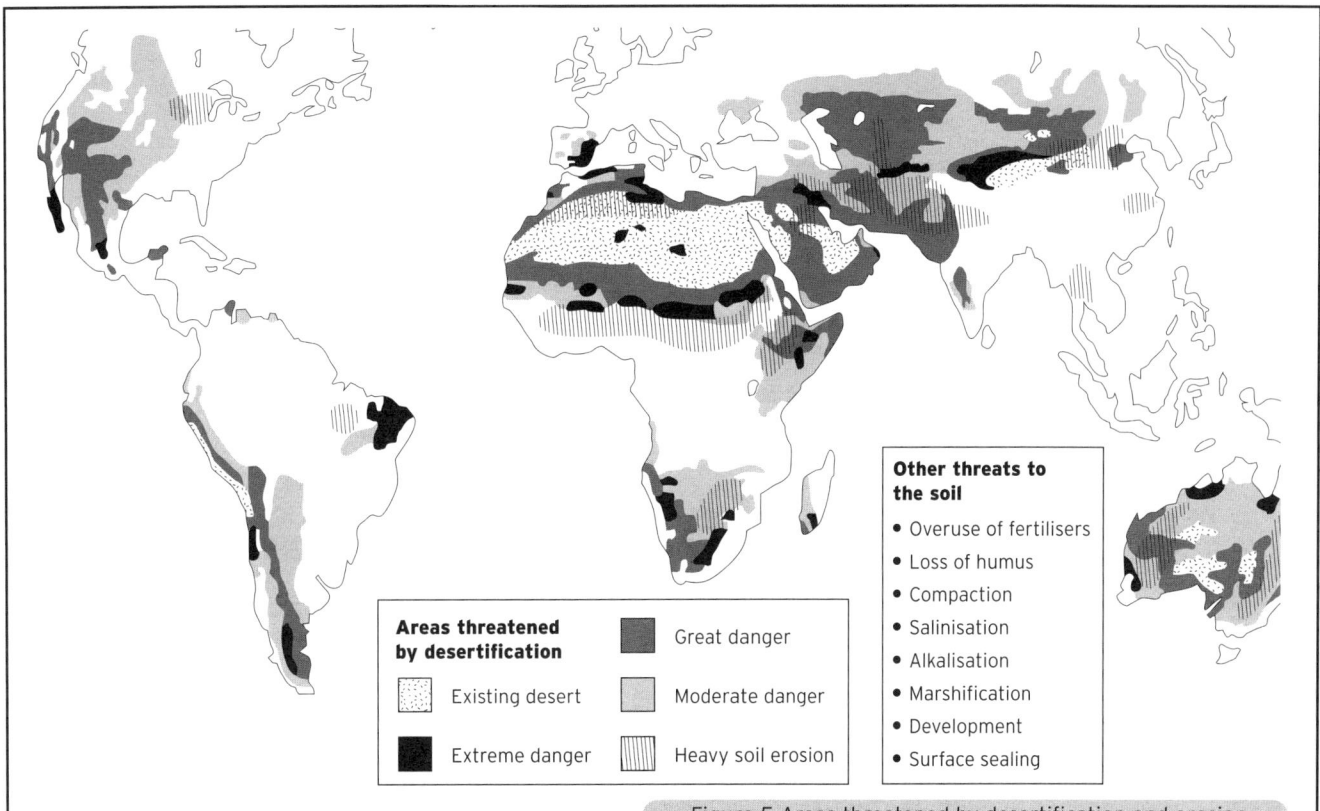

Figure 5 Areas threatened by desertification and erosion

Why is this happening?

Overgrazing and over-cropping combined with less fallow land causes desertification and results in falling productivity (Figure 4). There are droughts because there are no rains from ITCZ moving towards the tropics, so water sources dry up; farmers move to less suitable areas and the land is further degraded so they move again.

In the Sahel in Africa, clearing of natural vegetation for agriculture, especially cash crops such as ground nuts and cotton, is the main cause of this destruction. During the drought of the 1970s in the Sahel (meaning 'the shore') 1 million people fled from Burkina Faso and 250 000 died.

250 000 hectares are being lost each year in Niger through desertification. This is equivalent to 2 500 km² (an area about the same size as Luxembourg) a decade.

Solutions

The UN set the year 2000 as the original target for ending desertification in the Sahel in 1977. This has not happened, but working on a small scale has proved the most successful way of solving the problems. Solutions include:

- Reforestation
- Improved farming techniques
- Wind breaks
- Stone ridges to slow down surface flow and allow **percolation**
- Lessening dependence on cash crops to supply international demands or to service debt
- Use of irrigation but not large scale as it results in salinisation
- Aid such as Water Aid, Comic Relief, Cafod.

 Do you have examples of the threats to the soil listed on Figure 5?

1.8 The importance of marine and coastal ecosystems

Key questions for this section:

- Why is it important to safeguard the world's marine ecosystem?
- What are the pressures on coastal ecosystems?

You will be revising:

- Marine and coastal ecosystems
- The threats and solutions so far.

This is the third option in this unit. It is wider than coral reefs. If you live, for example, near the East Anglian or Hampshire and West Sussex coastal marshes, and have looked at marshland ecology as part of another A2 course, you might use it for your contrasting study.

Coral reefs and mangrove swamps

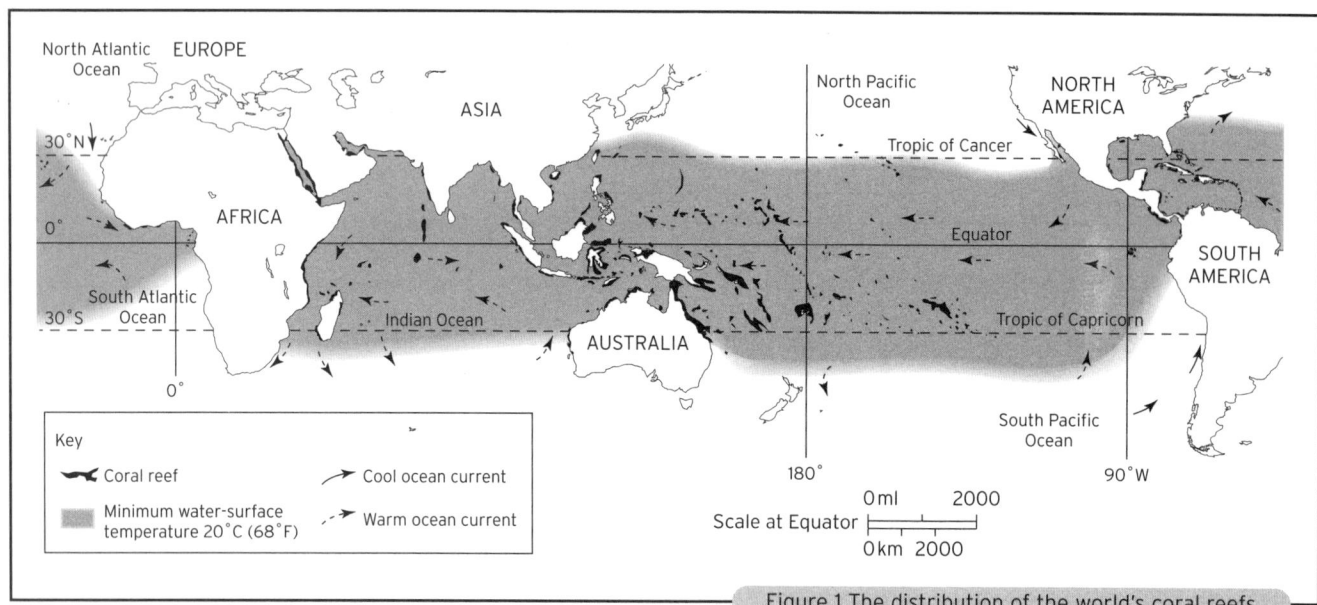

Figure 1 The distribution of the world's coral reefs

Where and why

Coral reefs are found between the Tropics in 0.2 per cent of the world's oceans. Indonesia and the Philippines have 20 per cent of the world's coral.
Coral thrives where there are:

- Shallow coastal waters – 30m maximum depth, and atolls
- Clear water which sunlight can penetrate
- The minimum water temperature is 20°C.

There is a **symbiotic relationship** between **coral** and **mangroves** (which filter out silt and pollution), and **sea grass beds**.

Characteristics

Coral reefs and mangrove swamps are one of the most delicate ecosystems on the planet because they form a living system that relies heavily on its living matter. The ecosystem is characterised by:

- Poor nutrient conditions – all nutrients are contained in living matter. There is negligible energy stored in litter or the 'soil'
- Coral polyps contain algae feeding on polyp waste that convert sunlight to energy

1.8 The importance of marine and coastal ecosystems

- Coral structures become homes and hiding places for an intricate food web for fish including predators
- Coral provides sand sediment in which mangroves and sea grass grow
- Once grown the mangroves then provide the spawning grounds, nurseries and foraging grounds for fish from the coral.

Functions

It is important to look after mangrove swamps and coral reefs because they:

- Provide a food source – 10–25 per cent of protein in South East Asia is from coral fish. Mangroves account for 30 per cent of Malaysian fish
- Protect coastal areas from erosion
- Provide the foundation for many islands, e.g. Aitutaki, Cook Islands
- Have high NPP (Net Primary Productivity) – higher than tropical rainforests.
- Attract tourists
- Mangroves can provide timber for fuel and building materials.

Threats

In Indonesia only 10 per cent of coral remains in prime condition; in the Philippines it is only 5 per cent. In total 58 per cent are threatened and 27 per cent are at high risk according to the World Resources Institute. However, 90 per cent of the reefs have not been assessed (see Figure 2).

Threats include:

- Climatic extremes – El Niño increases water temperature causing salinity changes, which leads to the loss of algae and coral loses colour, known as bleaching. Ninety per cent of Jamaican reefs are killed by hurricanes that smother reefs with algae
- Disease – sediment from storms blocks sunlight and causes diseases, e.g. black-band disease
- Accumulated environmental stress from diseases increases protective coating and uses up energy fighting viruses, bacteria and fungi and results in death of coral
- Inland pollution, i.e. sewage and agricultural runoff, increase plankton and crown of thorns sea stars that feed on coral – 22 per cent of all corals threatened
- Marine pollution – diesel spills, discharge of ballast, waste from boats – threatens 12 per cent of reefs
- Cyanide fishing for aquarium industry (not legal but fishing for aquaria is – worth $200 million)
- Food demands, e.g. for Groupers. Explosive fishing kills more than the fish collected for food. Overexploitation threatens 36 per cent of all corals
- Tourism – coral souvenirs, coral walking, diving. Coastal resorts destroying mangroves, e.g. Desaru, Malaysia
- Coastal development – resorts, ports, power station discharges, oil tanks and wells – affects 30 per cent of reefs. Also affects mangroves
- Fish farming destroys mangroves, e.g. Johore area of Malaysia.

Figure 2 Threats to coral reefs and mangrove swamps globally

What is being done?

- The Coral Reef Initiative (ICRI) was set up in 1995 to monitor reefs
- Government initiatives have been set up, e.g. in St Lucia. The Great Barrier Reef has been zoned – 20 per cent is now protected and 80 per cent managed for multiple use
- Marine parks have been created, e.g. seven islands off Johore.

Web link

For more research into marine and coastal ecosystems, go to the following website and enter the express code 1552S: www.heinemann.co.uk/hotlinks

Quick check

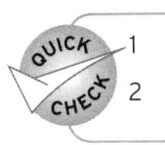

1. What are the functions of, and threats to, coastal marshes in the UK?
2. Draw a marine food web.

1.9 People and the future of ecosystem management

Key questions for this section:

- How are individuals and organisations playing an increasing part in ecosystem management?
- What is the future for global ecosystems?

You will be revising:

- Ecosystems and people
- Ecosystem conservation and conflicts
- Strategies for the future.

This section focuses on what people can do to at all levels from the United Nations right down to the individual. It is about taking a positive stance towards our future in spite of the problems we learn about in our studies (see Bob Digby's Global Challenges, *Chapters 9 and 10).*

Ecosystems and people

People as decision makers is a key element of sustainable development.

Non-governmental organisations

In the UK there is a range of non-governmental organisations (NGOs) campaigning for conservation. **Greenpeace** is probably the best known and it operates beyond national boundaries. **WWF (The World Fund for Nature)** is a larger scale NGO working for conservation. These and other organisations campaign for worldwide environmental causes, such as endangered species, and support activists working for the environment. At a more localised scale there are local environmental groups, some of which fight single causes, e.g. the A27 bypass for Arundel. It could be seen as a NIMBY organisation. Others represent the more radical opposition groups, e.g. Ecoaction.

National and local initiatives

Waste recycling is one of the most successful responses to Agenda 21 initiated at the Rio Earth Summit in 1992 (though the amount of waste doubled between 1995 and 2000). The UK lags behind much of Europe even though many UK authorities have introduced kerbside collections of green bins for rubbish that can be recycled (though plastics are restricted). In Germany, for example, recycling is much more common and better integrated: drinks cans are being outlawed in favour of recyclable and returnable bottles; bottle, paper and plastic banks are scattered through cities and cleared at very frequent intervals; at airports there are recycling containers at every waste bin site.

At a local level, Agenda 21 initiatives include the purchase by Hampshire Geographical Association of one acre of rain forest – geographers doing what they teach! They are thinking globally but acting locally.

Conservation and conflict

You should have studied two case studies for this topic. Here are two examples of case studies:

> **Vilcabamba, Peru**
> This area lies between two rivers in south central Peru on the fringe of the Andes and the Amazon basin. Up until now it has been protected by its isolated and remote position, but Conservation International now believe it is an area where development would threaten its unique environment. The area is home to two indigenous tribes, the Matisgenka and the Ashaninka. Some unique species have been found including a cloud forest rat known to have been an Inca pet. There may be many other species that have as yet to be identified. New settlers, who have been moving into the area to grow coffee, and possible pipelines, constructed by oil multinationals who are prospecting and drilling in nearby valleys, represent major threats.

> **Reminder**
>
> **NIMBY**
> Not in my back yard
>
> **Agenda 21**
> Part of the Rio Declaration on Environment and Development 1992 which dealt with 37 issues to encourage sustainable development. In the UK local authorities have been required to adopt Agenda 21 policies.
>
> **Sustainable Development**
> It has four principles: **Environment**, the preservation of the integrity of ecosystems; **Equity and social justice,** the granting of fair shares for all including the most disadvantaged locally and globally; **Futurity**, ensuring that present generations are able to leave future generations the ability to maintain standards of living; and **Public Participation,** making people aware of, and able to participate in, changes leading to sustainable development.

1.9 People and the future of ecosystem management

> **Dhani Forest, Orissa State, India**
>
> The forest's 2 200ha are used for food, fuel, building materials and natural medicines for the people in the five villages surrounding the forest. Orissa State permitted forest clearance by big companies in the 1970s. In the past the clearance of the forest to expand agriculture has resulted in less leaf mould to use as fertiliser. The cleared forest area was not as effective at enabling infiltration of water and lessened flow to wells used for drinking water and irrigation. By 1987 the villages had formed a forest protection committee which was recognised by the State. Other villages joined in the protection scheme. As a result, the forest canopy became denser until it was badly damaged by a hurricane in 1999.
>
> The villagers have legally recognised rights to use the forest although the State has greater rights. Many want the State to give up its rights to the people. The forest assists the local subsistence economy and, through the sale of timber, the local market economy. The costs of management are lower so saving Orissa State money. The villagers are restoring the forest but they still fear that the State may ignore local management. Success so far has depended on the local population

Strategic thinking

Some global strategies to meet concerns are outlined below.

- **United Nations Environment Programme (UNEP)** established 1972 in Stockholm. Priorities: 1) environmental health, 2) terrestrial ecosystems and development, 3) oceans, energy, and natural disasters. Operates EARTHWATCH – a system of surveillance intended to avert impending disasters which is working to encourage integrated action that includes governments and NGOs working together. Three *Global Environment Outlook* publications have led discussions.
- **Stockhom Declaration 1972**
- **World Conservation Strategy 1980**
- **Montreal Protocol 1989** responsible for initial control of gases harmful to the ozone layer.
- **Rio Earth Summit 1992** introduced Agenda 21 which was mainly adopted at a national and regional level, e.g. EU Action Plan towards sustainability 1993.
- **Millenium Summit 2000** where Kofi Annan, Secretary General of the UN, proposed *A sustainable future: the Environmental Agenda*

   As a part of the UN proposal, governments were urged to adopt a new ethic of conservation and stewardship of the Earth. It included:
   Climate Change: to adopt and ratify the Kyoto Protocol in 2002 as a step towards reducing emissions of greenhouse gases.
   Green Accounting: to consider incorporating the United Nations system of 'green accounting' into their own national accounts, in order to integrate environmental issues into mainstream economic policy.
   Ecosystem Assessment: to provide financial support for, and become actively engaged in, the Millennium Ecosystem Assessment, a major international collaborative effort to map the health of the planet.
- **Johannesburg Earth Summit, Rio +10** Confirmed the Kyoto principles.

> **Reminder**
>
> Try to have knowledge of at least two cases of local conflict in the conservation of areas. It is best to have one from another environment (Peru or Dhani) and one local and unique to you.

> **Reminder**
>
> Do you know Bruntland's definition of sustainability?

Biodiversity hotspots

Ecologist Norman Myers and **Conservation International** President, Ross Mittermeier, have identified **biodiversity hotspots** – 25 areas which still have 70 per cent or more of their endemic species. They occupy 1.4 per cent of the land surface and yet are home to 44 per cent of the Earth's plant species and 35 per cent of its fauna. Only about 40 per cent of these areas are protected and so the push is to extend protection to all hotspots (Figure 1).

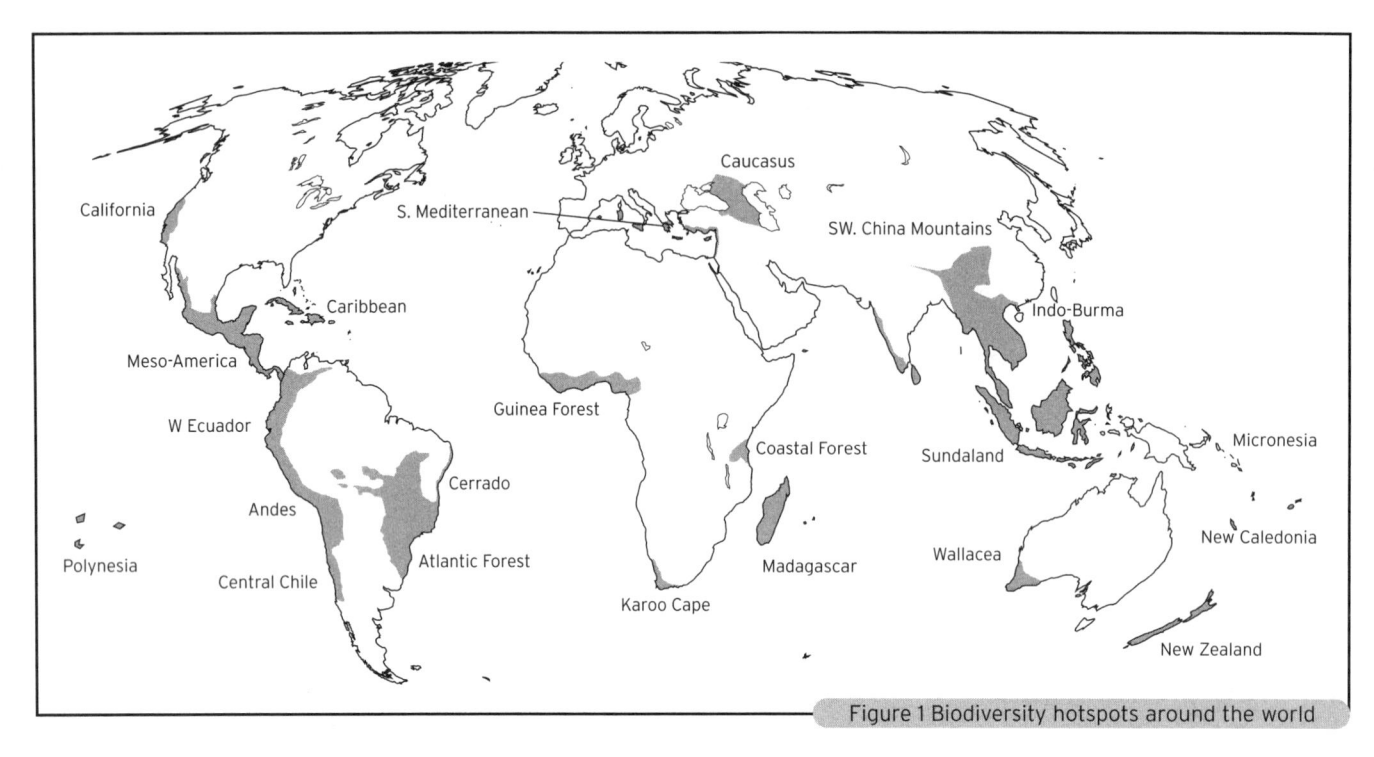

Figure 1 Biodiversity hotspots around the world

According to World Wildlife Fund these areas are also the most culturally diverse regions with 40 per cent of the world's ethno-linguistic groups living in them.

Also, 754 **World Heritage Sites** have been acknowledged as environments and ecosystems to be conserved (though not all are environmental – some are built environments e.g. Kew Gardens, London).

The future

Where is the greatest stress being placed on ecosystems? An Ecosystem Wellbeing Index has been constructed on the basis of:
- land quality and diversity
- the condition of inland waters and the sea
- air quality and contributions to atmospheric pollution
- the diversity of species and their maintenance
- and the use of energy and other extractive resources.

Web link

For more information on ecosystems and their management, go to the following website and enter the express code 1552S: www.heinemann.co.uk/hotlinks

QUICK CHECK — For any one hotspot make notes on the reasons why it has been classed as a hotspot. For more information, go to the web link.

1.10 Measuring population change

Key questions for this section:
- What are the components of population change?
- How is population change measured and portrayed?

You will be revising:
- Methods of portraying and measuring populations
- Population change over time.

Population or, more precisely, **demography** (the empirical study of population data) is the first topic that contributes to Section B in Unit 4 of the examination. You are expected to answer one question from (Population and the Economy) (Bob Digby's Global Challenges Resource File, Heinemann 2002 contains many case studies that you can use).

Population is the flow of people through time and comprises four elements as shown in Figure 1.

Portraying population

Much of our interpretation of density depends on the scale of the units used to show density. Figure 2 shows population density in England and Wales. Figure 3 shows the increase in densities of population for India's States and Districts, 1991–2001. These are **choropleth maps**.

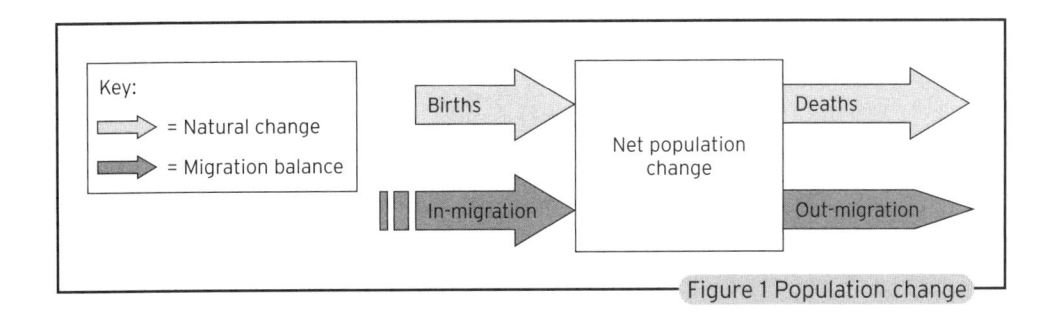

Figure 1 Population change

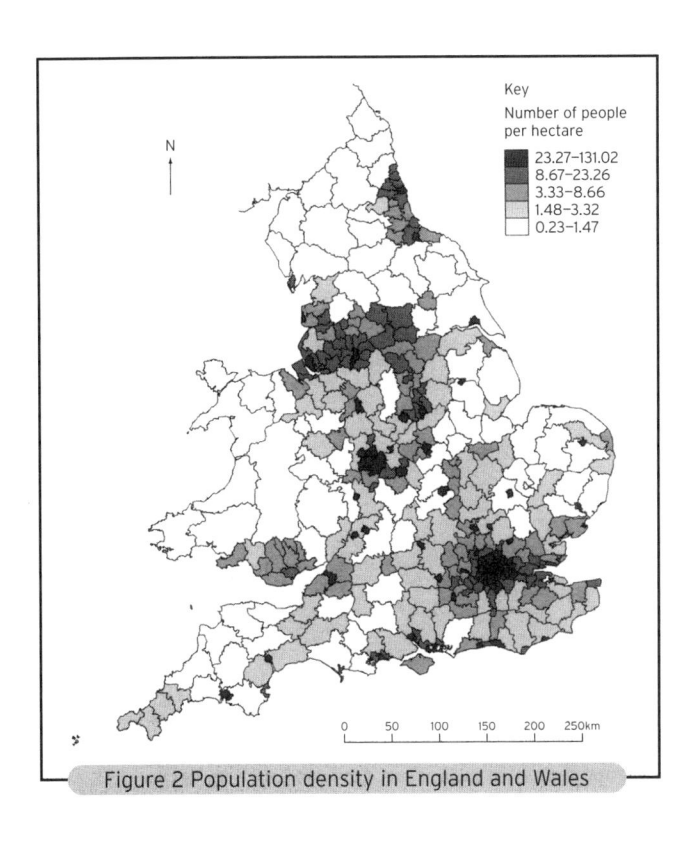

Figure 2 Population density in England and Wales

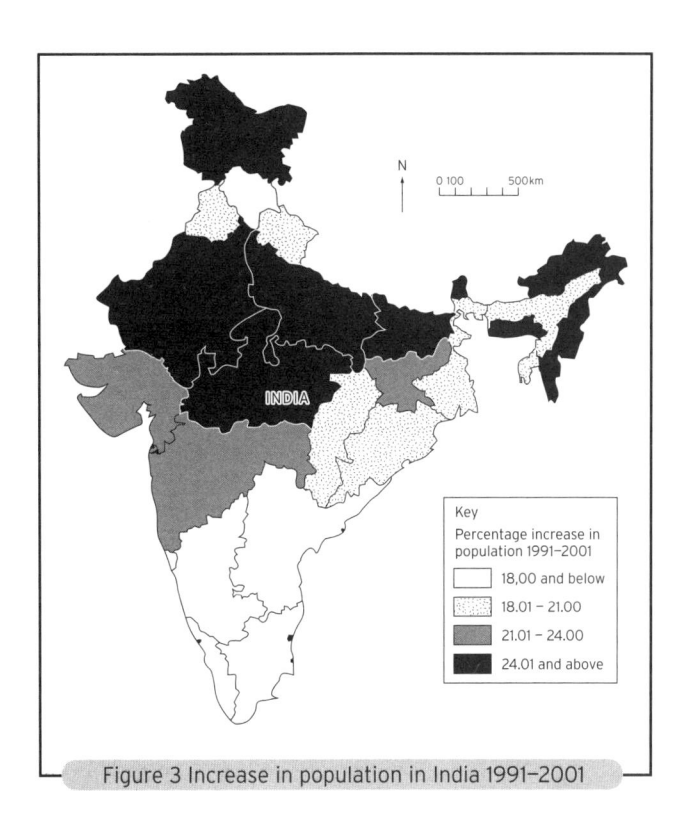

Figure 3 Increase in population in India 1991–2001

Measuring population change

How do we measure population change?

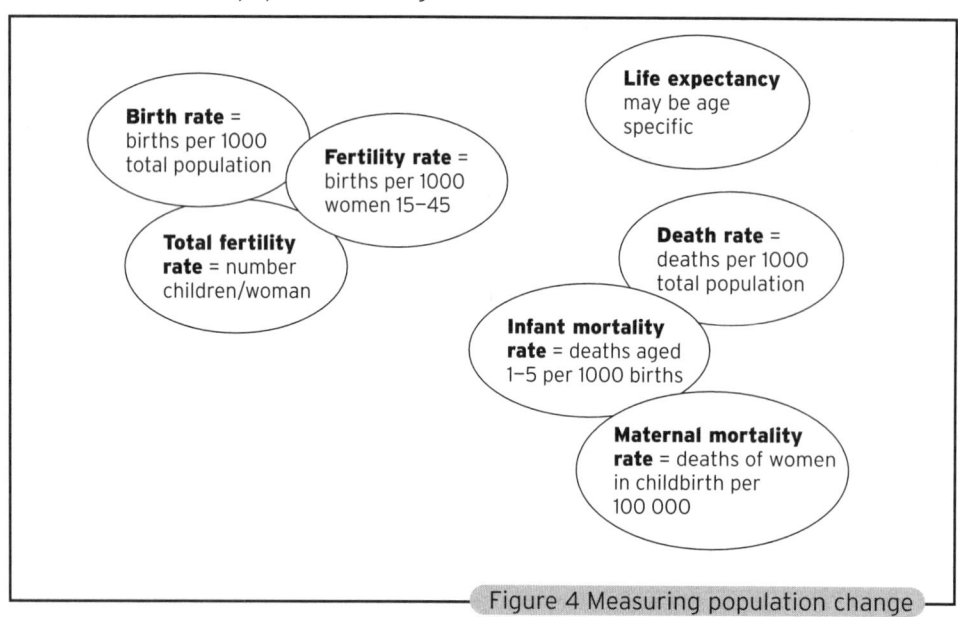

- **Birth rate** = births per 1000 total population
- **Fertility rate** = births per 1000 women 15–45
- **Total fertility rate** = number children/woman
- **Life expectancy** may be age specific
- **Death rate** = deaths per 1000 total population
- **Infant mortality rate** = deaths aged 1–5 per 1000 births
- **Maternal mortality rate** = deaths of women in childbirth per 100 000

Figure 4 Measuring population change

Population pyramids

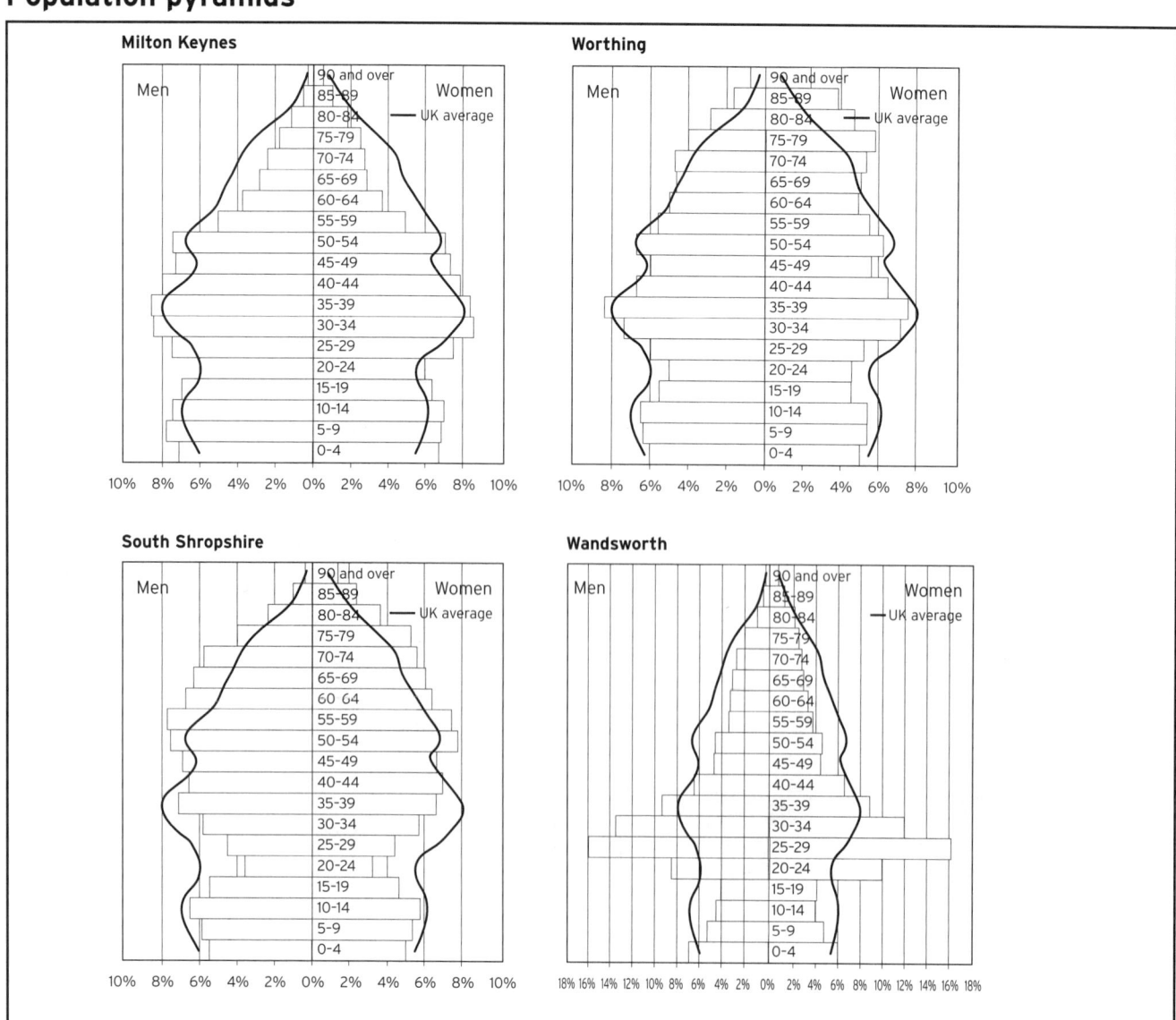

Figure 5 2001 Census Pyramids for Milton Keynes, South Shropshire, Wandsworth and Worthing

1.10 Measuring population change

Figure 6 shows that fewer people are being born in the UK, which will have the long-term effect of creating an ageing society. There will be fewer people entering the labour market and there might be labour shortages in the future. In-migration, invariably young adults, can be seen as the **demographic replacement** for the lower fertility rate.

Fertility rates are an indicator of development. Table 1 includes demographic data for a selection of countries that illustrate levels of development. Can you identify levels of development from the table?

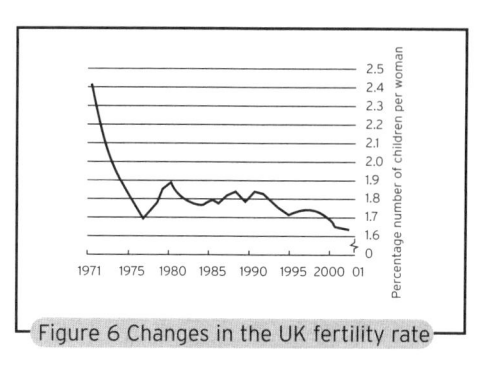

Figure 6 Changes in the UK fertility rate

	Population (millions)	Birth rate (per 1000 pop.)	Children per woman (1999)	Infant mortality (per 1000 live births)	Life expectancy	Death rate (per 1000 pop.)
Chad	8.4	48	6.0	122	45	19
Mali	11.1	50	6.5	130	51	18
Bolivia	8.7	33	4.3	66	61	9
Zambia	10.7	44	5.5	94	41	20.7
Saudi Arabia	21.6	35	5.7	25	71	4.4
UAE	2.7	16	3.4	12	75	3.6
S Korea	49.5	15	1.7	5	73	6
Singapore	4.1	14	1.7	5	77	5
Poland	38.5	11	1.5	10	73	10
Hungary	9.8	9	1.4	10	71	5
Canada	30.8	11	1.6	6	79	7
Germany	82.1	9	1.3	5	77	10
Italy	57.6	9	1.2	5	77	9
UK	58.8	12	1.7	6	M75 F80	11

Source: www.unfpa.org, the UN Population Fund website

Table 1 Population data for a selection of countries 2001

Figure 7 shows how you can explain many of the variations of demographic indicators.

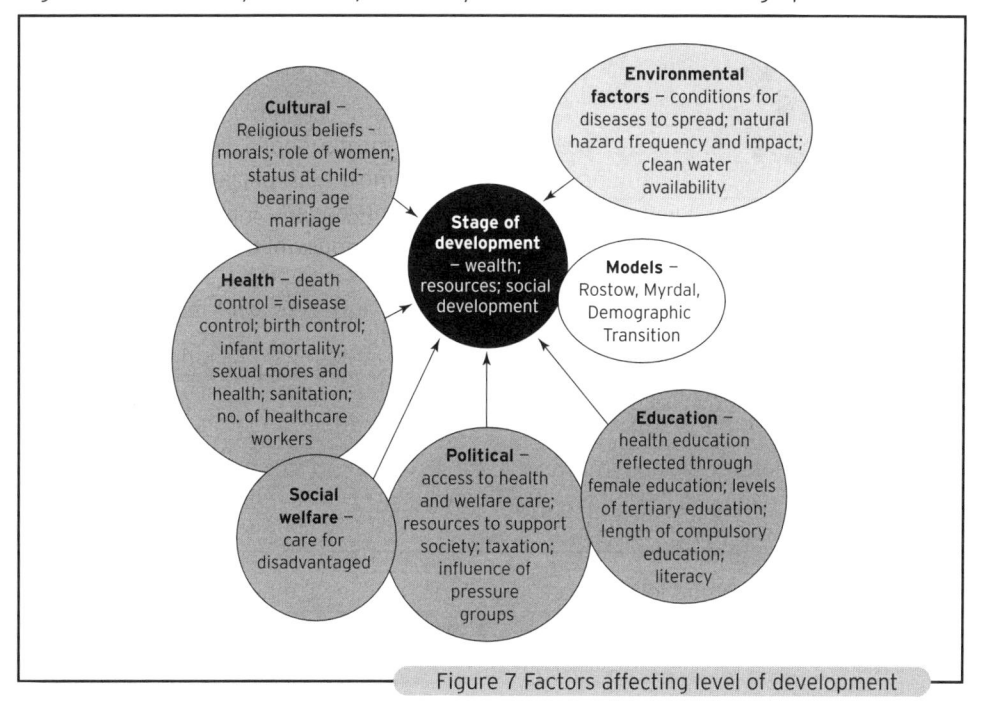

Figure 7 Factors affecting level of development

- **Cultural** – Religious beliefs – morals; role of women; status at child-bearing age marriage
- **Environmental factors** – conditions for diseases to spread; natural hazard frequency and impact; clean water availability
- **Health** – death control = disease control; birth control; infant mortality; sexual mores and health; sanitation; no. of healthcare workers
- **Stage of development** – wealth; resources; social development
- **Models** – Rostow, Myrdal, Demographic Transition
- **Social welfare** – care for disadvantaged
- **Political** – access to health and welfare care; resources to support society; taxation; influence of pressure groups
- **Education** – health education reflected through female education; levels of tertiary education; length of compulsory education; literacy

Global challenge

Population change through time

The five stage **Demographic Transition Model** (Figure 8) is a popular way to model population changes over time. It was devised in 1929 by Warren Thompson and needed updating for the twenty-first century. The fifth stage, population decline or negative growth, first proposed by demographers such as R. Cliquet is an essential addition to the model.

This model must however, be viewed within the context of its conception. It only outlines the changes as they occurred in Europe in the nineteenth and twentieth centuries. Many factors now affecting population change mean that few LEDCs are likely to follow the same patterns today.

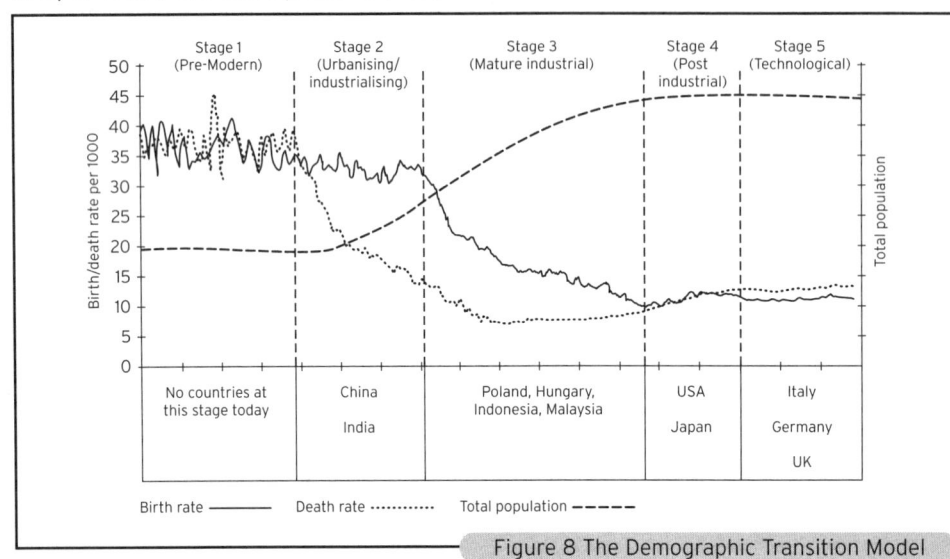

> **Reminder**
>
> This is a model and not all countries will experience all of the stages.

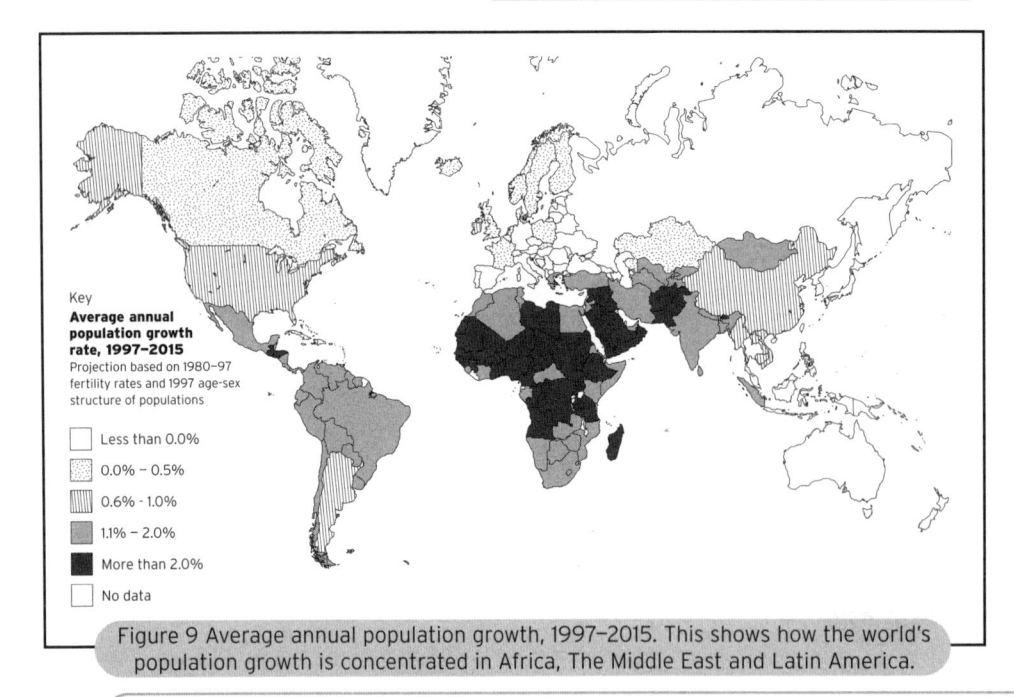

Figure 8 The Demographic Transition Model

Figure 9 Average annual population growth, 1997–2015. This shows how the world's population growth is concentrated in Africa, The Middle East and Latin America.

QUICK CHECK

1. What other type of map (other than Figure 2) can show population distributions?
2. Do you know how the measures of natural change in Figure 4 are indicators of development?
3. What explanations are there for the variations in population structure shown by the pyramids in Figure 5?
4. Can you show how the **levels of development** discussed in 1.13 can be found in the data in Table 1 (1.13)? Are there anomalies and how can these be explained?
5. Using the Demographic Transition Model as a guide in Figure 8, attempt to explain the population growth patterns shown in Figures 2 and 3.

1.11 The national and international challenges of population change

Key questions for this section:

- What are the global challenges posed by population change?
- What are the national challenges posed by population change?
- How do decision makers manage population change?

You will be revising:

- The challenge of global population growth
- Measuring population change
- The role of the census
- Population and resources.

The challenge of population growth has been around for 200 years and today there are ever more difficult problems to solve that impact on all of our lives.

Global population growth

Most growth is occurring in LEDCs (98 per cent), so much so that only 1 billion of the 6 billion people on this Earth live in MEDCs. India's population is increasing by 2.1 per cent a year. At this rate it will double itself every 35 years. The situation may become unsustainable in less than 25 years when India's population would be 1500 million. If further resources can be used to support the population then unsustainability may be delayed or even prevented.

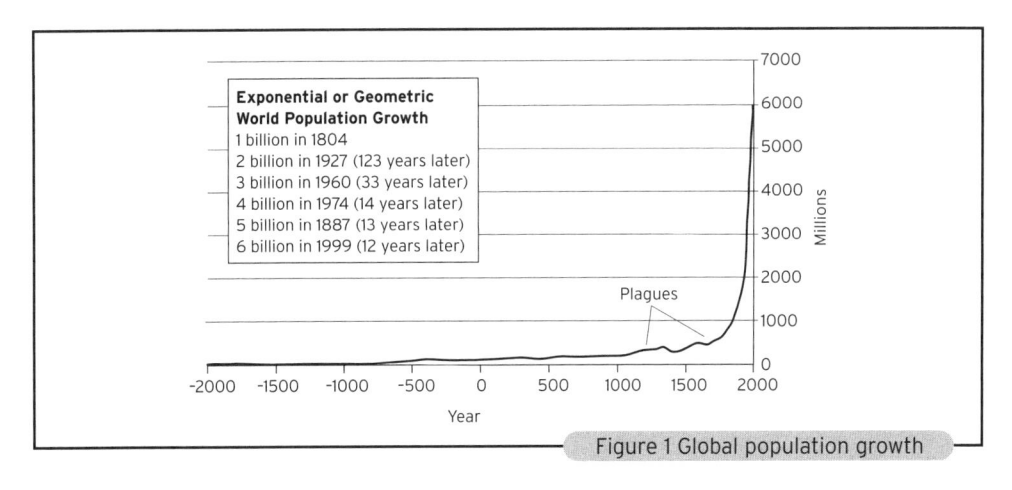

Exponential or Geometric World Population Growth
- 1 billion in 1804
- 2 billion in 1927 (123 years later)
- 3 billion in 1960 (33 years later)
- 4 billion in 1974 (14 years later)
- 5 billion in 1887 (13 years later)
- 6 billion in 1999 (12 years later)

Figure 1 Global population growth

Reminder

Some important concepts include:

- **Sustainable population (or optimum population)** – the number of people who can be supported without depleting its resources while maximising output as the economic, technological and social conditions of a country advance. This is a purely theoretical concept which varies according to a country's level of development.
- **Overpopulation** – when the resources are insufficient to sustain a growing population without a reduction in living standards. Famine, disease and war may reduce overpopulation.
- **Underpopulation** – a theoretical concept that suggests there are countries whose populations are unable to optimally exploit their resources because they lack the population to do so. The population could have higher living standards if it could exploit its resources.

Population and resources

There is a desired balance between population and resources. A population needs food, minerals and energy as well as human resources such as brain power and inventiveness to support itself and enable it to develop in a sustainable fashion.

Population theories

1 Thomas Malthus 1798

- Population, if unchecked, grows geometrically, i.e. 2, 4, 8, 16, 32, 64 etc
- Food demand increases arithmetically, i.e. 1, 2, 3, 4, 5, 6 etc
- With less food for everyone more people die from starvation – **a positive check**
- Other checks include reduced fertility due to: 'moral restraint'; 'vice', i.e. abortion, infanticide and contraception,; further positive checks, e.g. war and disease
- Population growth declines
- Two variables only – food and population. Possible examples of Malthusian population collapses due to declining food supplies include the droughts in Sahel, Ethiopia and Zimbabwe. Malthus's predictions have only occurred in some countries and NOT globally because food supply technology, biotechnology and transport have improved.

2 Esther Boserup 1965, Findlay 1982

Boserup suggested that population growth, rather than being a hindrance to economic growth, is actually a prerequisite for development. In her model the population–resource ratio is modified by out-migration.

Findlay suggested that over-population is moderated by trade in goods, ideas and technologies.

More recent contributors suggest that overpopulation in developing countries is one consequence of, and prerequisite for, their economic underdevelopment.

3 Neo-Malthusians – P. Ehrlich, The Population Bomb, 1968, The Club of Rome, Limits to Growth 1972

Results from increased demand for food and resources:

1. Either a) food and resources run out
   Or b) policies are developed to regulate growth
2. Either a) death rate rises and population declines
   Or b) industrialisation occurs and fertility decreases

Whether the course is 1) or 2) population growth is reduced.

Rapid population growth hinders development, raises the **dependency ratio**, reduces the amount of national income that might be available for investing in economic growth and causes hunger. Many of these characteristics can be produced by mismanagement of the economic system. Rapid depletion of non-renewable resources, deterioration of the environment, ecological disasters, rising tensions and violence in the world may lead to overpopulation. In LEDCs, population growth will outstrip the countries' and the entire world's capacity to provide food and other resources. The loss of a resource base will hamper economic growth and may account for ecological crises such as **soil erosion**, **flash floods** and **salinisation** of agricultural lands due to poor management.

Coping with population change

Examples of projected population increase include:

- Ethiopia – 62 millions in 2000 will rise to 213 millions by 2050.
- Pakistan – 148 millions will rise to 357 millions.
- Nigeria – 122 millions will rise to 339 millions.

Demographic fatigue is the reaction to high fertility and growth. It results in declining numbers. The impact of AIDS on Africa is one such case. The population of Zambia has tripled since 1964, but twenty per cent of Zambia's population is HIV positive and they are from the productive age groups. Twenty per cent of the population will die from AIDS by 2010.

Zambia's population policy, adopted in 1989, is being revised to include the impact of HIV/AIDS on population, the 2000 national gender policy, the health and development of adolescents, and the Zambia Poverty Reduction Strategy. The population growth rate and fertility rate both remain high. Recognising the links between population and maternal and child health, Zambia's government, with donor assistance, is supporting family planning, maternal-child health programmes, with some focused on youth. The emphasis on preventing HIV infection, caring for those affected, and introducing anti-retroviral (ARV) treatments is overshadowing attention on population growth.

Combating high growth rates

The objective of the **International Conference on Population Development (ICPD)** held in Cairo in 1994 was to improve the rights for females, especially in areas of education and reproductive health, while also noting demographic gains from lower child mortality and declining fertility. It aimed to create reproductive programmes that built on existing family rights.

> **Reminder**
>
> Bob Digby's *Global Challenges* has case studies of Ethiopia and Colorado and many good background maps on the issue of population change.

1.11 The national and international challenges of population change

1 China

China has 20 per cent of the world's population. It has many policies to curb population growth. In the late 1970s the State Birth Planning Commission was founded and its rigid and controversial methods assisted in bringing the fertility rate from 5.9 in the 1960s to 1.8 in 2000. Methods included a one child policy; mother to wait until of age to begin a family; special incentives for one-child families, including cash payments, free schooling, and better employment opportunities; penalties such as compulsory sterilisation, return to the state of financial benefits, and pressure to have an abortion, were enforced upon people who had more than two children. Methods raised concerns about human rights violations and China was criticised by the international community. Now 32 areas allow couples to have as many children as they want and the government is trying to change attitudes about families rather than enforcing strict policies.

> **Reminder**
>
> Do develop other case studies of policies from differing countries.

LEDC policy solutions – female rights

Cultural and social issues, human rights, the environment, and economic and political changes are linked to rapid population growth. In LEDCs, women may lack self-esteem, employment, schooling and the full legal and social rights of citizenship. They depend on children for status and security. Therefore policies to reduce population include:

- giving all girls a good education attempting to empower women outside the home
- providing employment or small business opportunities to women
- enabling women to shed feelings of dependence in patriarchal societies which traditionally expect large numbers of children
- educating people at reproductive health clinics about the value of women's lives
- promoting marriages at later ages and emphasising contraceptive methods for men as well as women
- influencing African tribal leaders to promote lower birth rates.

In addition:
- The United Nations puts pressure on countries to control population rates backed up by aid.
- Some countries have migration policies to less crowded areas, e.g. **Transmigration**.
- There is spontaneous migration to leave overcrowded countries, e.g. illegal migration Mexico to USA, Filipinos to Japan, UK.

Ageing populations

UK life expectancy was 45 for men and 49 for women in 1901. In 2001 it was 75 and 80 respectively for children born that year. Men aged 65 in 2001 could expect to live until they were 81 whereas women aged 65 could expect to live until they were 84. The elderly are going to outnumber the young.

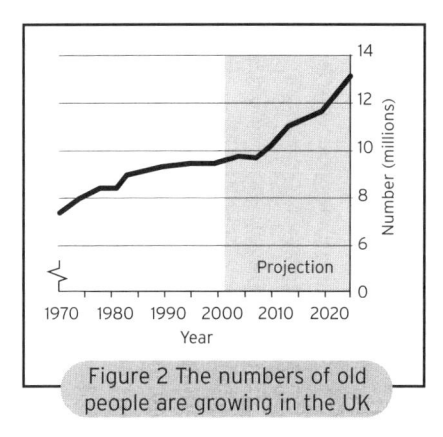

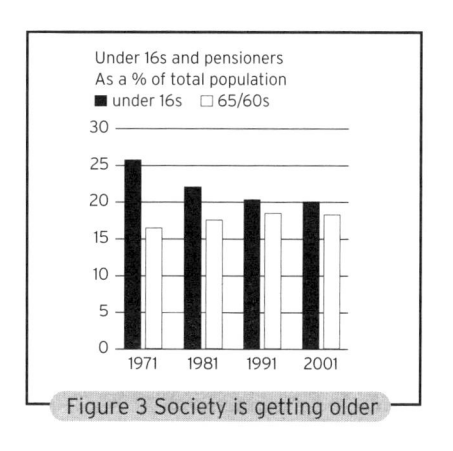

Figure 2 The numbers of old people are growing in the UK

Figure 3 Society is getting older

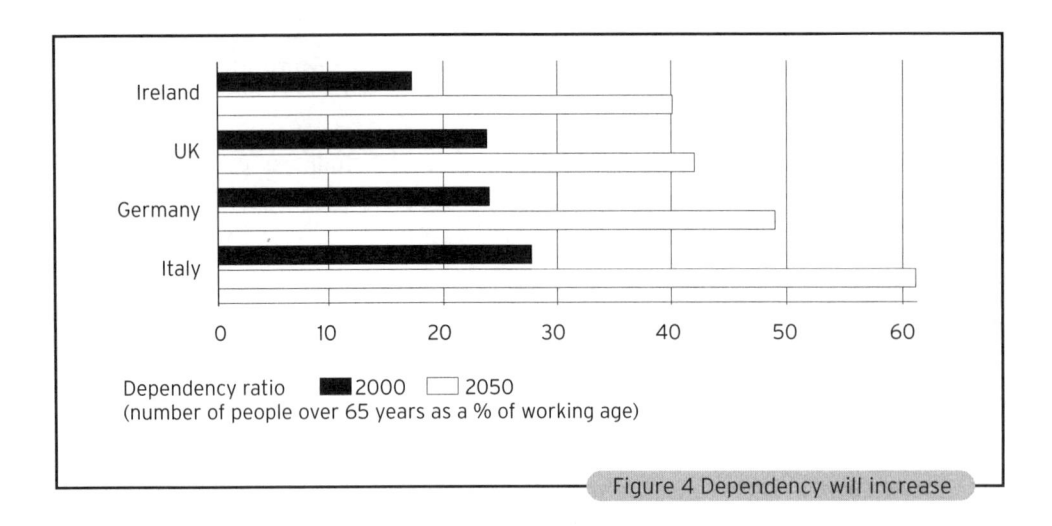

1.11 The national and international challenges of population change

Figure 4 Dependency will increase

Dependency ratio ■ 2000 □ 2050
(number of people over 65 years as a % of working age)

- Ireland: ~17 (2000), ~40 (2050)
- UK: ~24 (2000), ~42 (2050)
- Germany: ~24 (2000), ~49 (2050)
- Italy: ~28 (2000), ~61 (2050)

What are the effects of an ageing society?

Figure 5 The effects of an ageing society

Ageing society:
1. Shortage of workers
2. Immigration encouraged
3. Work until older
4. Pensions crisis
5. 'Grey' vote – political power
6. Birth incentives
7. Demand for public transport
8. More nurses, fewer teachers
9. Effects on health care
10. Spending power – retailing
11. Increased tourism
12. Recreation demands – keep fit

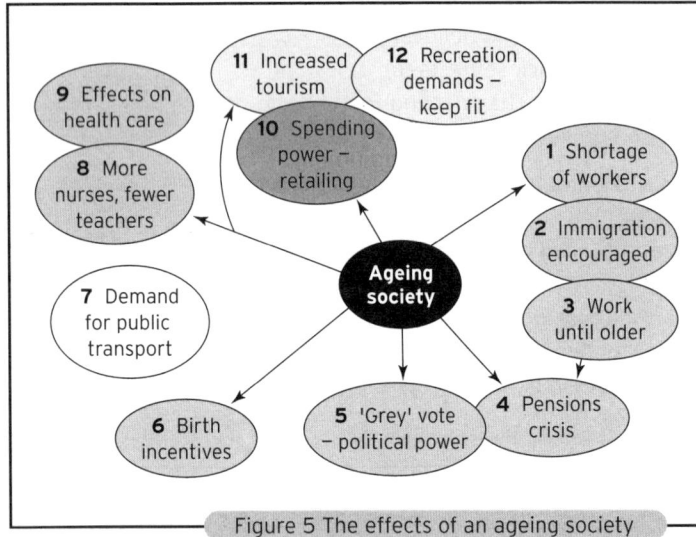

1. Labour shortages can be tackled in MEDCs by introducing automation or robots – not possible in LEDCs due to lack of investment
2. Immigration – a hot topic due to surge of nationalism and protectionism; USA favours selective immigration
3. Employing elderly – a solution of many retail firms, e.g. in the DIY sector
4. Changing savings rules – made more difficult by poor investment climate
5. Political parties are wooing the older voters
6. Birth incentives are in place in Russia
7. Demand for public transport will rise for very old, e.g. London's free pass for over sixties is encouraging a switch away from cars earlier in life
8. Government investment in training switching to caring sector. Recruiting from LEDCs a temporary solution for nursing but affects nursing in LEDCs
9. Demand for nursing homes; bed blocking in hospitals a problem
10. Those with private or occupational pensions have spending power – no mortages – ability to select areas of spending
11. Retirement travel and second home purchases rising – often leads to retirement migration; evidence of return retirement migration when a partner dies.
12. Recreational clubs (not just bingo!): ramblers, keep fit, running.

QUICK CHECK

1. An ageing society will affect LEDCs. What are the effects on LEDCs?
2. Assess a range of national population policies saying what is good and acceptable, and what might be unacceptable.
3. With reference to one country, outline the benefits and costs of an ageing society.

1.12 International migration

Key questions for this section:

- **What are the causes and impacts of international migration?**
- **What are the key issues posed by international migration?**

You will be revising:

- Types of international migration
- Models of migration
- Issues and impacts.

Migration is one of the most current global challenges. In studying international migration you must make sure that you do not confuse propaganda with accurate information.

Types of international migration

Migration is a complex phenomenon involving the decisions of many people. Migration affects the areas being left and the areas to which migrants move. More people migrate and more global issues of wealth and poverty are being revealed by migration.

Push factors
Population pressure; poor quality of life; economic hardship – lack jobs; ethnic cleansing; persecution; disasters; harsh environment; war; marriage; retirement.

Pull factors
Economic opportunities; perceived living standards; society perceived as receptive; job opportunities; partner in work in country; seen as receptive to refugees and asylum seekers.

Return migration 'Didn't work out', perceived benefits not found, expelled, prejudice

Figure 1 Causes of international migration: push/pull factors

Type	One year	Seven years	Temporary	Life	Return possibility
Job	Yes	Yes	?	?	Possibly
House/Marriage	Yes	Yes	?	?	Changed circumstances
Expatriot	Yes	Yes	Normally	No	Yes – contracted
Retirement	Yes	Yes	Possibly	Yes	Later in life
Refugee	Yes	May be forced to return	May return	Yes	If circumstances in origin change
Asylum	Yes	Yes	Yes	Possibly	If rejected
Emigration	Unlikely	Yes	Unlikely	Yes	Unlikely

There is a possibility that there can be temporary international migration for marriage and jobs

Table 1 Duration of migration

Forced migrations

Examples include:

1. Ethnic Germans from Moldova to Germany at the end of World War II – part of a larger scale movement of millions in 1945–6.
2. Rwanda to République Démocratique du Congo following genocide of Hutus.
3. Ethnic cleansing in the former Yugoslavia.
4. Palestinians into West Bank and Gaza (up to 3.2 million).
5. Colombia – clashes with guerrillas forced up to 1 million to migrate; much internal but some international.
6. Asylum seekers – Kurds to UK, Afghans to Australia.

> **Reminder**
>
> Terms for migrants:
>
> **Economic migrants/migrants**
> People who move to another country and may return if they wish.
>
> **Refugees**
> People who flee their country because of a well-founded fear of persecution for reasons of race, religion, nationality, political opinion or membership of a particular social group.
>
> **Returnees**
> Migrants or refugees who return voluntarily.
>
> **Voluntary repatriates**
> People who are assisted by UNHCR to return home.
>
> **Asylum seekers**
> People who have left their own country because of persecution and are seeking a safe haven. They may be granted refugee status and allowed to stay.
>
> **Internally displaced**
> Victims of a civil war who have left their home for another part of the country.
>
> **Stateless persons**
> People with whom the right of citizenship is in dispute, e.g. Roma peoples in Eastern Europe.

> **Reminder**
>
> Do not confuse **migration** with **movement**. Movement is short term such as commuting, holidays and going to university.

Voluntary migration

1 Labour migration or economic migrants: Italians in 1950s to Bedford brickworks; temporary contract workers, e.g. Portuguese in Channel Islands, Poles in London hotels, Latvian fruit pickers, Mexicans to California.
2 Skilled labour migration: Financiers to New York and from USA in London, Filipino and South African nurses to NHS in UK.
3 Voluntary refugees following disasters: Montserrat (but returning).

Laws and models of migration

Ravenstein's Laws of Migration were statements which predicted the usual pattern of migration. They are described on page 170 in Bob Digby's *Global Challenges*.

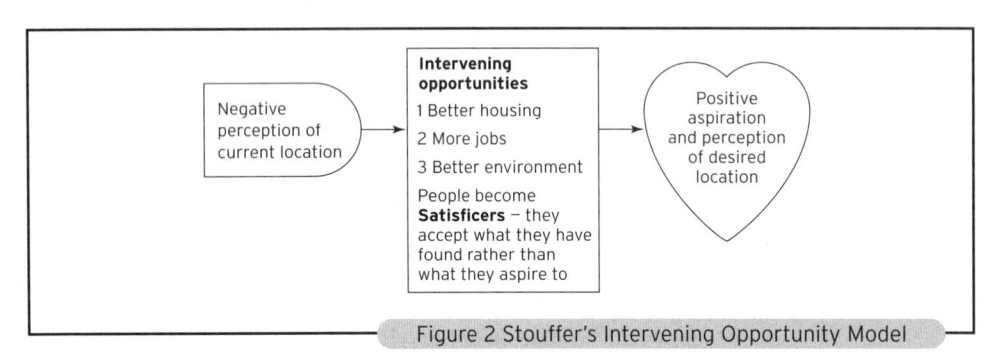

```
Negative                Intervening              Positive
perception of           opportunities            aspiration
current location   -->  1 Better housing    -->  and perception
                        2 More jobs              of desired
                        3 Better environment     location
                        People become
                        Satisficers – they
                        accept what they have
                        found rather than
                        what they aspire to
```

Figure 2 Stouffer's Intervening Opportunity Model

The **Push–Pull Model** (Figure 1) is another basic model based on factors that pull the immigrant to the destination and those that push him/her away from the country of origin.

Lee's model is a variant which examines people's:

- perception of a destination
- assessment of their existing circumstances
- assessment of the costs of migration which includes the distance to be moved
- personal circumstances before they make a decision.

Todaro's model is based on two conclustions:

- Economic factors most important of push–pull factors
- Migration likely where urban incomes greater than rural ones.

All these models assume an individual decision maker whereas for many it is a group or family decision.

FACTORS ENCOURAGING A MOVE	SEARCHING WHERE TO GO	OBSTACLES TO MOVING	DECISION, STAY OR MOVE
Ability to move Finance and income Age factors Gender factors Education levels Family ties Personality **Stresses** Poor housing Poor environment Lack of employment Persecution Intolerance Poverty Population pressure Colonial ties	**Information on** Alternatives Housing Jobs Social facilities Family present Other migrants Willingness to accept refugees and asylum seekers **Where to get info.** Acquaintances Family Media Government Agents (legal and illegal).	**Family** Relatives not willing to move **Distance** Physical distance Perceived distance Cost distance **Barriers** Selection conditions Qualifying conditions Quotas **Reality** Entirely different	Based on the balance of answers to questions in previous 3 columns

Table 2 Decision factors for migrants

1.12 International migration

	BENEFITS	COSTS
Demographic	Natural increase slows because young adults leave	Population becomes older
Resources	Slightly less pressure on existing resources	Fewer people to produce food
	Skills from returnees to develop resources	
Economic	Remittances sent back	Loss of skilled workers
	Return migrants bring skills and money to invest	Less opportunity for country to develop
	Pensions coming in from retired migrants	
Social	Tolerance	Westernisation of returnees
	Solves social and racial tensions	Cultural imperialism
		Diminished cultural diversity

Table 3 Costs and benefits of international migration on country of origin of migrants *SOURCE*

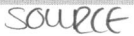

	BENEFITS	COSTS
Demographic	Demographic replacement for declining populations with low negative change	White flight – migration of original population
		Gender concentrations, e.g. where only males migrate
Resources	Human resources enhanced	Pressure on resources might grow unless contribution to economy raises living standards
Economic	Labour force needs filled (temporary or permanent)	Social support if unemployed, educational demands
	Unwanted jobs by host society filled	Illegal labour on poor, illegal wages – bad reputation
	Higher pay for skilled workers who then pay taxes	Low pay for unskilled
Social	Multicultural society – religions, languages, food, retailing, music	Areas dominated by group – ghettoisation and segregation
		Racial tension

Table 4 Costs and benefits of international migration on country receiving migrants *HOST*

Issues and impacts of migration

Refugees and asylum seekers

The UN projection is that between 2000 and 2050, the MEDC population of 1.2 billion people in 39 countries will not change. However, the 4.9 billion LEDC population will rise to 8.2 billion in the same period. The population of Europe will decline by 5 per cent by 2025 and therefore the labour force will have to be replaced either by a higher rate of natural increase or by immigration of economic migrants, refugees and asylum seekers.

One in 20 persons is forced to leave their country. At present there are 22 million refuges according to UNHCR and most are women and children.

There have been refugees throughout history. They include refugees from:

- post-war Europe
- India and Pakistan in 1947
- Cuba in the Castro era
- Vietnam 1975 (800 000 Vietnamese Boat People of whom 2000 are still in Hong Kong)

Reminder

You should have examples for key questions about costs and benefits.

- Myanmar 1991–2 (250 000 Muslims of whom 21 000 still in Bangladesh)
- Sri Lanka 1992–5 (800 000 Tamils of whom 64 000 are still in India).

Today, Kosovo, Sierra Leone, Angola, Rwanda, Burundi, Somalia, Sudan and Guatemala all have problems that are giving rise to refugee movements.

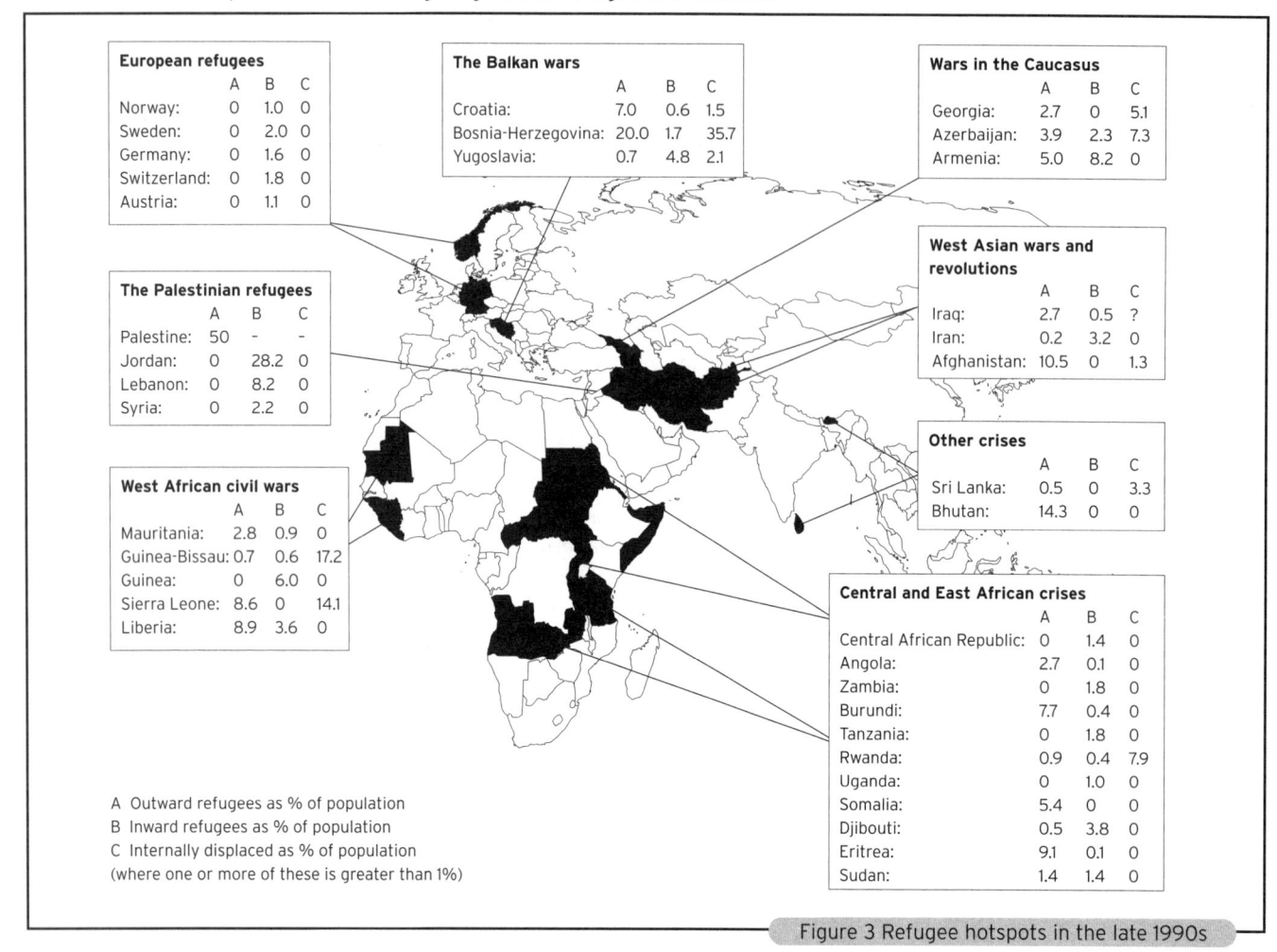

European refugees

	A	B	C
Norway:	0	1.0	0
Sweden:	0	2.0	0
Germany:	0	1.6	0
Switzerland:	0	1.8	0
Austria:	0	1.1	0

The Balkan wars

	A	B	C
Croatia:	7.0	0.6	1.5
Bosnia-Herzegovina:	20.0	1.7	35.7
Yugoslavia:	0.7	4.8	2.1

Wars in the Caucasus

	A	B	C
Georgia:	2.7	0	5.1
Azerbaijan:	3.9	2.3	7.3
Armenia:	5.0	8.2	0

The Palestinian refugees

	A	B	C
Palestine:	50	–	–
Jordan:	0	28.2	0
Lebanon:	0	8.2	0
Syria:	0	2.2	0

West Asian wars and revolutions

	A	B	C
Iraq:	2.7	0.5	?
Iran:	0.2	3.2	0
Afghanistan:	10.5	0	1.3

West African civil wars

	A	B	C
Mauritania:	2.8	0.9	0
Guinea-Bissau:	0.7	0.6	17.2
Guinea:	0	6.0	0
Sierra Leone:	8.6	0	14.1
Liberia:	8.9	3.6	0

Other crises

	A	B	C
Sri Lanka:	0.5	0	3.3
Bhutan:	14.3	0	0

Central and East African crises

	A	B	C
Central African Republic:	0	1.4	0
Angola:	2.7	0.1	0
Zambia:	0	1.8	0
Burundi:	7.7	0.4	0
Tanzania:	0	1.8	0
Rwanda:	0.9	0.4	7.9
Uganda:	0	1.0	0
Somalia:	5.4	0	0
Djibouti:	0.5	3.8	0
Eritrea:	9.1	0.1	0
Sudan:	1.4	1.4	0

A Outward refugees as % of population
B Inward refugees as % of population
C Internally displaced as % of population
(where one or more of these is greater than 1%)

Figure 3 Refugee hotspots in the late 1990s

Figure 3 shows that refugees and asylum seekers are an issue in many countries. In Europe, foreign-born population as a percentage of the population is higher in Germany, Switzerland, Austria, Sweden and Luxembourg than in the UK. In Germany asylum applications are 100 000 per year, about the same level as the UK. There were 45 000 applications in 1999 to the Netherlands and 22 000 to Belgium. Attempted illegal crossings into Germany in 1999 were 40 000.

Reactions

Throughout Europe there have been many reactions to the influx of refugees.

- 1999 Tampere, framework for a common European policy
- 2002 Seville, EU summit discussed but little progress.
- 2002 German government recognised that economic migration essential as demographic replacement for declining population. Need for 50 000 per year but the emphasis will be on the highly qualified.
- Political opposition:
  - Austrian Freedom Party – Jorg Haider
  - Jean-Marie Le Pen, France, National Front, 'France for the French'
  - Vlaams Blok in Belgium, Philip de Winter, 'our own people first'
  - Netherlands, List Fortuyn, 35 per cent vote in Rotterdam, anti Islam
  - Switzerland 2003, People's Party becomes the largest in parliament, Christoph Blocher 'foreigners are welome in Switzerland so long as they are not illegal immigrants'
  - National Front in UK.

1.12 International migration

- Various policies to deal with migrants at source (but not practical for asylum seekers)
- Pre-boarding arrangements imposed on airlines
- Acceptance of those with the shortage skills
- Prevention measures – boat patrols, border fences and security
- Repatriation and carrier pays
- Safe third country as holding place for those without visas – Germany and east European states
- Temporary accommodation until safe to return – Bosnians in 1996
- Policies which address the issues at source through aid to help eliminate poverty – Sweden uses this policy.

Immigration to the UK

In 2001, 8 per cent of the UK population was born outside the country and included 1 per cent Ireland, 1 per cent European Union. Of those who declared ethnic status, 2 per cent were Black or Black British, 5 per cent Asian or British Asian and 4 per cent UK non-white. The UK has 4 million migrants out of a population of 58 million.

Immigration in 2001 was a net inflow of 172 000 people. Asylum seekers peaked in 2001 and in 2003 were down to 4000 per month compared with 16 000 per month in 2001–2. This was due to a 70 per cent drop in applications from Iraqis. Numbers can easily be misinterpreted for political reasons.

Issues

- Media portrayal of immigration – Sangatte and the Channel Tunnel, emphasis on tragedies. Use of emotive language by right wing press such as 'flood', 'tidal flow' and 'swamping' services
- Difficulty of distinguishing asylum seekers from economic migrants
- The work of people traffickers
- Communities in UK encouraging others to join them – leads to chain migration This is the most common reason for immigration
- Continuing push to migrate from areas of political and social unrest
- A 1 per cent rise in UK incomes will result in a 0.67 per cent rise in immigration.

Benefits of accepting refugees and asylum seekers to the UK

These are very much the same as for international migration (see Table 4). In addition:

- More work permits have been issued to fill labour gaps under **Highly Skilled Migrant Programme** – 200 000 in 2003.
- There is also the humanitarian reputation of a country which is enhanced when refugees are accepted.

Costs

- Pressure on housing, social services and education provision. There is a tendency for refugees to settle in the core of the country and not the periphery, e.g. London, Paris, Milan, Munich
- Welfare payments (though these might be small if a person finds employment)
- Cost of controlling ports of entry by Home Office
- Costs of policing opposition when it is expressed in violent, racist attacks – Wrexham 2003, Burnley 2002
- Dispersal costs
- Repatriation costs.

Figure 4 Distribution of foreign born in England and Wales in 2001

QUICK CHECK

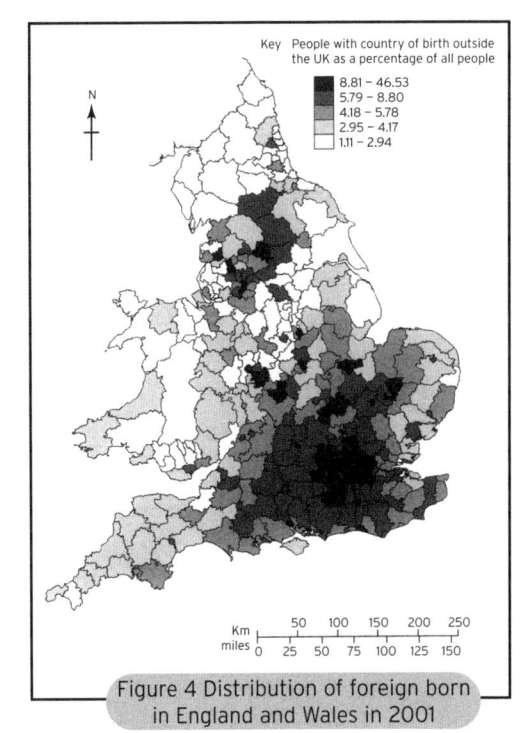

1. What are the benefits of accepting refugees and asylum seekers?
2. What factors explain the distribution of foreign born in Figure 4?
3. The UK has only the eleventh largest number of immigrant workers. Why might the USA, Russia, Germany, France, India and the Ukraine all have more? Will the issues in these countries be the same as in the UK?

1.13 Global economic groupings

Key questions for this section:
- What are the main global economic groupings?
- What links exist between groupings?
- What are the factors encouraging the global economy?

You will be revising:
- The global economic groupings
- The North–South divide
- Globalisation and global interdependence
- The development gap
- Aid and trade.

This is a key section because it introduces the underlying concepts and theories in modern economic geography. The subject matter is the way people earn their living and the ways in which this is changing all over the world. It is also about a shrinking world where it only takes 24 hours to reach any part of the globe and where people are dependent on the work of others far away. This is a true Global Challenge.

> **Reminder**
>
> **The development continuum**
> The world has a gradation of economies ranging from the very poorest to the richest. Simplistic divisions make a useful model but mask the reality of a continuum.
>
> **The development gap**
> The widening gulf between the world's richest and poorest nations is increasing and measures to close the gap are not agreed. It has replaced the **North–South divide** because it places greater emphasis on closing the gap.

The global economic groupings

You are probably used to a threefold world: MEDCs, LEDCs and NICs. Most economists recognise the fact that there is a continuum of levels of development and that the groupings are just a convenience for labelling.

Table 1 shows some characteristics of types of economic states.

	G8		MEDC		NIC/RIC		Oil rich		Former Soviet		LEDC		LDC/HIPC	
	USA	UK	NL	Spain	Singapore	S. Korea	Saudi Arabia	UAE	Poland	Czech Republic	Bangladesh	Ghana	Burundi	Ethiopia
GNI	31,910	23,590	25,140	14,800	24,150	8,490	6,900	ND	4,070	5,020	370	400	120	100
GNP/cap	29,240	21,410	24,780	14,100	30,170	8,600	6910	17,870	3,910	5,150	350	390	140	100
Urbanisation	77%	90%	90%	78%	100%	64%	87%	90%	63%	75%	26%	36%	9%	16%
HDI rank 2000	7-0.94	11-0.93	5-0.94	18-0.91	28-0.88	30-0.88	73-0.76	48-0.81	35-0.83	32-0.85	145-0.48	129-0.55	171-0,31	168-0.33
GDI 2001	0.94	0.93	0.93	0.91	0.88	0.87	0.74	0.80	0.84	0.85	ND	0.56	0.33	ND

GNI = **Gross National Income** the sum added by all producers, plus taxes and income from abroad in $US.
HDI = **Human Development Index** An index measuring average achievement in three basic dimensions of human development: a long and healthy life, knowledge and a decent standard of living. The ranking is from best to worst.
GDI = **Gender-related Development Index** measures the three dimensions of the HDI adjusted to account for inequalities between men and women. This has the effect of reducing the values.
LDC = **Least Developed Country**
HIPC = **Heavily Indebted Poor Countries**

Table 1 The major groups of economies

1.13 Global economic groupings

LDCs or Least Developed Countries (Figure 1) have been defined by the UN.

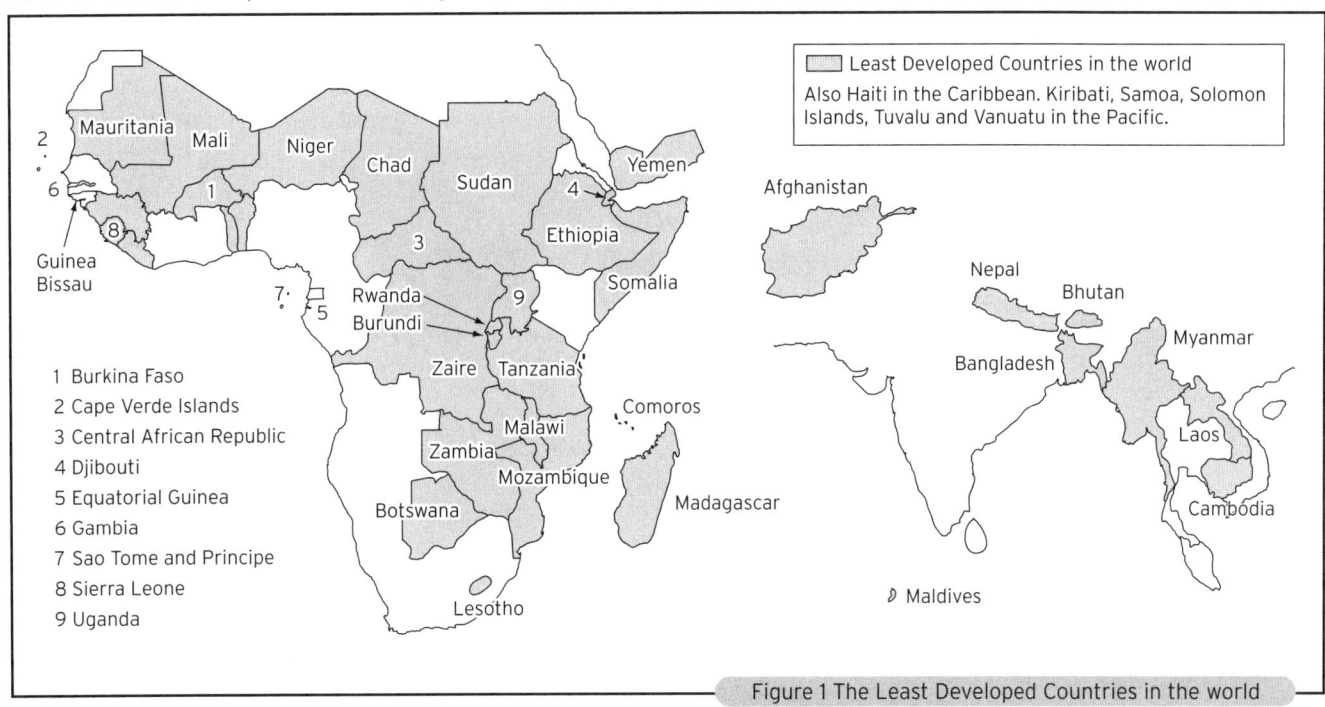

Least Developed Countries in the world
Also Haiti in the Caribbean. Kiribati, Samoa, Solomon Islands, Tuvalu and Vanuatu in the Pacific.

1 Burkina Faso
2 Cape Verde Islands
3 Central African Republic
4 Djibouti
5 Equatorial Guinea
6 Gambia
7 Sao Tome and Principe
8 Sierra Leone
9 Uganda

Figure 1 The Least Developed Countries in the world

International economic groups

Other groups that you will find mentioned whose statistics might be useful:

- **G10 or Paris Club**
- **G15 group** of 17 developing countries from Asia, Africa and Latin America, set up to foster cooperation and lobby the WTO and G8
- **G77** now comprising 135 members, promotes South-South cooperation
- **OPEC** represents the oil trade interests of oil producers (but not Norway and UK), and campaigns for interests of LEDCs
- **ASEAN** is a regional trade alliance for SE Asia
- **FTAA** is the Free Trade Area of the Americas founded in 2001 to promote all of the Americas' trade
- **The European Union (EU)** was founded in 1957 by the Treaty of Rome as the Common Market, a single market for goods and services. It will grow to 25 members in 2004 when 10 other countries join the existing 15 members. It is the trade bloc which governs UK trading and is the UK's largest market.
- **Andean Community** was founded in 2003 to foster development and is an example of a regional bloc brought together by their common environment.
- **World Trade Organisation** replaced GATT in 1995 and is the governing body for world trade and has 144 members
- **OECD** has 25 members who all cooperate to encourage economic expansion of their members
- **IMF** (the International Monetary Fund) is, alongside the World Bank, the organisation that assists the world's economic development. It is often seen as a lackey of TNCs (Transnational Companies) and the USA.

Web link

For research into these groups, go to the following website and enter the express code 1552S:
www.heinemann.co.uk/hotlinks

The North–South Divide: The uneven spread of wealth

In 2003, Mohamed Bennouna, the Moroccan Chairman of the economic group for South–South, G77, expressed serious concern over the widening gap between developed and developing nations. G77 expressed its concern regarding the marginalisation of a large number of developing countries, which are being denied the benefits of development, particularly in Africa.

This summarised press release merely reflects the opinions of the Independent Commission on International Development, chaired by Willy Brandt, whose findings,

Figure 2 The G15 summit

Figure 3 The World Bank view of the world's economies

Key
- Low-income economies
- Middle-income economies
- Upper middle income economies
- High-income economies

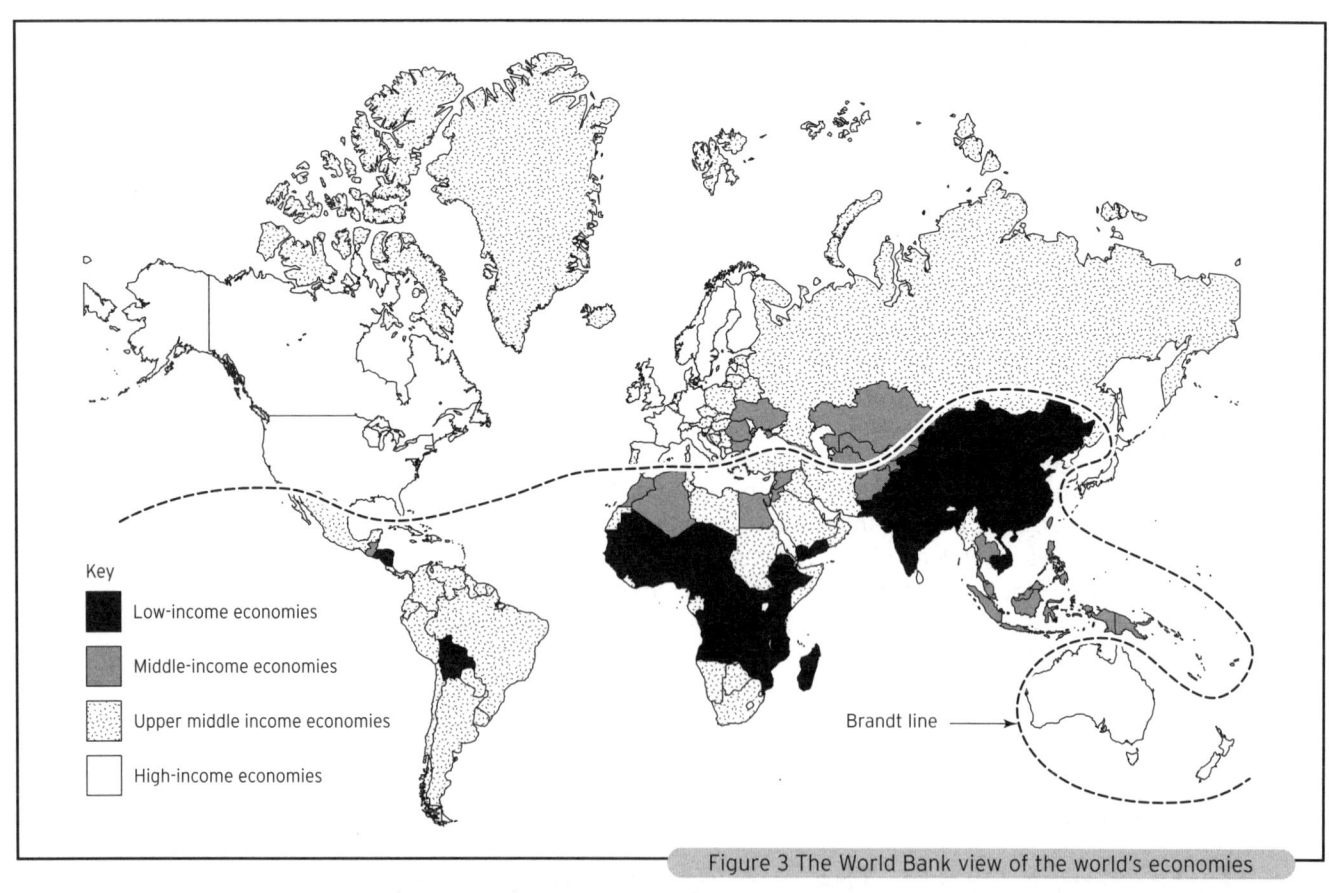

Brandt line

North–South: A Program for Survival, were published in 1980. It called for urgent action to bridge the gap between the rich **North** and the poor **South**. It recommended levels of aid equivalent to 1 per cent of GSP (Gross Social Product – the money spent on housing, literacy and health care; a more Marxist definition of wealth than GDP), greater access to low interest loans, agricultural development and the stabilisation of food supplies and prices, more support for the South to market its own raw materials rather than depend on TNCs and, finally, a raft of policies to enable the South to have access to energy resources. Little progress has been made as G77 have indicated.

1.13 Global economic groupings

The development gap

This and the following sections link to 1.16, pp 66–68, How might the development gap be closed?

Figure 4 illustrates the development gap by showing how GDP is obtained in a threefold world.

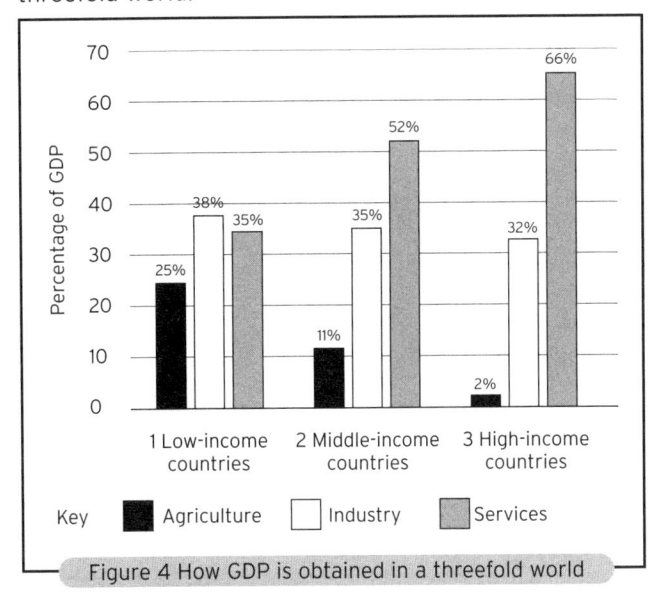

Figure 4 How GDP is obtained in a threefold world

- Low-income countries: Agriculture 25%, Industry 38%, Services 35%
- Middle-income countries: Agriculture 11%, Industry 35%, Services 52%
- High-income countries: Agriculture 2%, Industry 32%, Services 66%

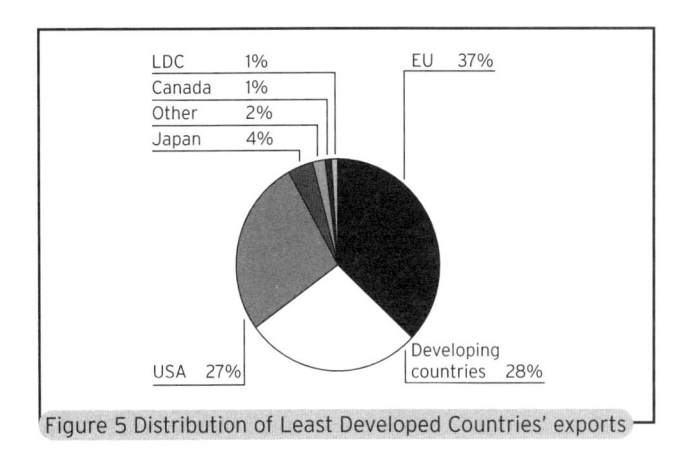

Figure 5 Distribution of Least Developed Countries' exports

- EU 37%
- Developing countries 28%
- USA 27%
- Japan 4%
- Other 2%
- Canada 1%
- LDC 1%

The global economy

Trade is still skewed towards the Developed World as Figure 5 shows for the Least Developed Countries. Generally:

- The basis of trade is in primary commodities and manufactured articles which accounts for 80 per cent of world trade. The rapidly growing service trade (often called invisibles) still only accounts for 20 per cent.
- Trade growth is strongest among the better-off middle-income economies such as Brazil, Mexico, Poland, Hungary and the oil states.
- Some low income economies have grown rapidly in the past decade, e.g. Vietnam.
- The poorest 48 economies have seen their trade remain static.
- Over 76 per cent of world trade takes place between high income countries because they are trading manufactured goods and high value added products. 23 per cent of trade originates from the rest of the world.

	Exports $ billions	% Exports	Imports $ billions	% Imports
Food	24.0	7.2	44.5	12
Agricultural raw materials	4.6	1.4	12.5	3.4
Ores and non-ferrous metals	5.2	1.6	19.6	5.3
Fuels	5.7	1.7	56.4	15.2
Crude petroleum	0.5	0.1	40.0	10.8
Petroleum products	4.7	1.4	8.2	2.2
Manufactured goods	282.2	85.0	237.8	64.0
Chemicals	42.5	12.8	14.2	3.8
Machinery and transport equipment	151.7	45.7	92.9	25.0
Other goods	10.5	3.2	0.6	0.2
Total	332.1	100	371.4	100

Table 2 Trade flows between the European Union and low and middle income economies (1999)

- Trade emanating from Sub-Saharan Africa is only 1.3 per cent of the world total and that from South Asia (India, Pakistan, Sri Lanka and Bangladesh) is a mere 1.1 per cent of global trade.

Table 1 shows the trade flows of the European Union with the World Bank's low and middle income economies in 1999. Note the manufactured goods trade surplus, the import of foodstuffs and agricultural raw materials e.g. cotton and wool, and the import of raw materials.

Part of the answer to the reminder is the fact that the prices of the primary commodities that are often the life blood of LEDCs have declined in relative terms since 1990 as the table below illustrates for 2000.

> **Reminder**
>
> What does the data in tables 1 and 2 tell you about the low income of the LEDCs?

Commodity	Price Index 1990 =100	Commodity	Price 1990	Price 2000
Food	84	Cotton (cents/kg)	182	129
Agricultural raw materials	91	Rubber (cents/kg)	86	68
Fertilisers	105	Cocoa (cents/kg)	127	90
Metals and minerals	82	Tea (cents/kg)	206	186
Petroleum	122	Coal ($/metric tonne)	39.69	26.01
Steel products	76	Natural gas ($/mmbtu)	2.55	3.82
Cotton	80	Petroleum ($/bbl)	22.88	27.97

Table 3 Prices of primary commodities 1990–2000

In all cases the prices of raw materials have gone down which affects the raw material exporters who have to export more to retain the same income. The cocoa grower in Ghana and the tea pickers in India are the losers as are mine workers in Zambia. Only petroleum and gas exporters have gained.

Aid

There are many types of aid. You should at least be familiar with the list below.

- **Food Aid** and **Disaster Aid** are very immediate responses to need. Up to 12 per cent of bilateral aid still comes in this form. Sometimes food aid is longer term.
- **Bilateral Aid** flows from one country to another to finance projects normally of interest to the donors because it will depend on the expertise that they supply. Therefore much of the money returns to the donor country. About three quarters of aid is in this form.
- **Multilateral Aid** is aid from a variety of sources that is normally managed by global agencies such as the World Bank.
- **Program Aid** supports key specified imports when there is no money to purchase them.
- **Project Aid** is tied to large-scale projects such as many of the large dam schemes, Pergau, Akosombo, Aswan etc.
- **Tied Aid** is used to tie the recipient to the donor country especially when the donor wishes to gain a specific advantage. **Untied Aid** leaves the recipient with the choice.
- **Debt relief** is a form of aid because it cancels the outstanding interest payments. It is used to reduce the debt burden of countries that, while repaying loans, are seeing the outstanding sums grow despite repayments.

Figure 6 Aid from Development Aid Committee (DAC) members

Aid donors are normally ranked on the amount donated as a percentage of GNP. The UN recommendation is for the equivalent of 1 per cent of GNP. The graph below (Figure 6) shows that many rich countries are failing to deliver.

Development assistance to the poorest economies has declined in the past 15 years. This is partly due to economic crises in the donor countries such as Japan. Today

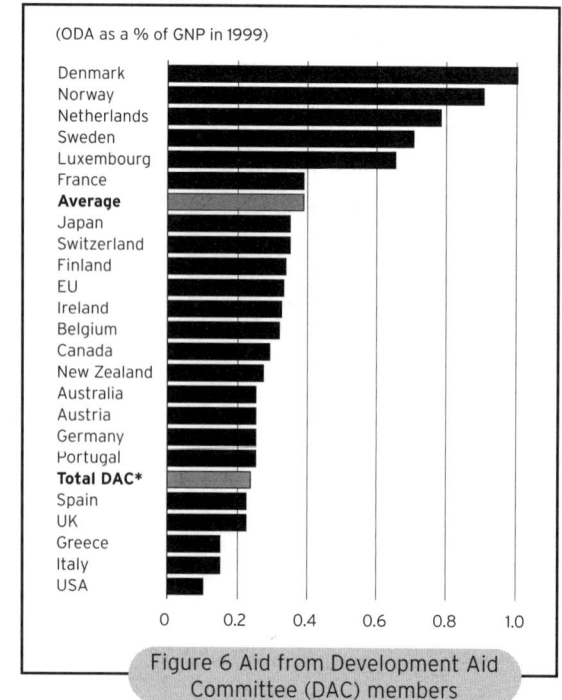

1.13 Global economic groupings

Foreign Direct Investment (FDI) exceeds development assistance. However, the poorest nations do not have access to this funding because they are regarded as bad risks by the international money markets.

Globalisation and interdependence

Global interdependence – This term is used to suggest that all economies depend on goods and services from one another. It was the antidote to **Dependency Theory** which explained why LEDCs remained poor. It suggests that trade and aid are dominated by MEDCs and TNCs in a form of **neo-colonialism** whereby they gain an unfair trading advantage. Interdependence puts a positive spin on the relationship suggesting that our lifestyles and economic wealth are interlinked with those of primary producers, for example agro-businesses produce flowers and runner beans all year round in Kenya. They benefit both the Kenyan and UK economies.

Emerging from interdependence is the concept of **sustainable development**: 'improving the quality of life for all of the world's people without increasing the use of our natural resources beyond the Earth's carrying capacity'(UN 2002). The quote goes on to say, ' It may require different actions in every region on the world, but the efforts to build a truly sustainable way of life requires the integration of action in three key areas: economic growth and equity, conserving natural resources and the environment, social development'.

This is the global challenge agenda that comes from accepting interdependence and the reality of globalisation.

Globalisation – This is a much abused term. For geographers it has three subdivisions:

1. **Political globalisation**, the burgeoning influence of western democracies and especially the USA. It has been aided by the post-1989 collapse of the USSR and the 9–11 New York attack. It is the oldest form of globalisation because the British Empire, the French Empire and even the Roman Empire were all attempts to politically dominate the known world.
2. **Cultural globalisation** is the western cultural influence over the arts (cinema is often seen as the first truly global activity), media and music.
3. **Economic globalisation** is the growing control of the economies of countries by a range of **Trans-national Companies (TNCs)** whose HQs are based in the G8 countries. It is flow of money rather than flow of goods.

What has aided economic globalisation?
- The rapid growth and spread of information technology and computing power enabling ideas, information and money to move rapidly
- Declining real costs of transport and communications, e.g. air fares, containerisation.
- Free market economics espoused by WTO and especially USA where especially the Republican Party's electoral funds are mainly supplied by TNCs
- The role of WTO and trade blocs such as the EU promoting free trade
- The growth of the service economy – insurance, tourism, share dealing, futures markets and derivatives
- The effects of bilateral aid on countries having to repay that assistance
- The growth of the TNCs who are responsible for most trade by value.
- Growth of English as the language of business.

What are the challenging issues?
- Environmental degradation when exploiting primary resources
- Potential for pollution due to lax laws and law breaking
- The increasing divide between those working for the new activities and the rest
- Migration to the key production regions without adequate planning for new housing
- The westernisation of employees
- The employment of women in societies where it is not traditional for women to seek work
- The use of foreign, ex-patriot labour in the managerial functions.

There is also the effect of globalisation on the developed countries.

> **Reminder**
>
> You need to have a case study developed of the impact of globalisation on a NIC or RIC (Newly/ Recently Industrialised Country) such as Malaysia, Singapore, Taiwan and South Korea. The example should contain details of the types of inward investments, the growth of exports, places where production is taking place and an assessment of the costs to and benefits for the economy, environment and people.

> **Quick check**
>
> 1. Is there any difference between dependency and global interdependence?
> 2. Summarise some of the effects of globalisation on the economy, environment and society of a country in West Europe such as the UK.

1.14 Changes in the location of economic activity

Key questions for this section:
- What changes are taking place in the character of the global economy?
- Who are the key players in changing the location of economic activity?
- What are the implications of the changes in economic activity?

You will be revising:
- The causes and effects of the global shift in economic activities
- Deindustrialisation and TNCs
- The role of government.

This section examines the key challenge of globalisation and global shift. It is an enormous topic that is also covered in Bob Digby's Global Challenges.

Global shift in economic activities

This is the term used to describe the changing geography of production and services in the world brought about by the growth of new activities and the patterns of investment and government policies towards economic growth. It involves the shift of production and some services to low cost locations. The consequences may be **deindustrialisation** in MEDCs and industrialisation in NICs, RICs and LEDCs. Globalisation has environmental effects, for example by spreading and increasing the sources of carbon dioxide.

As countries develop, the proportions employed in the sectors of the economy change. However, the model (Figure 1) is an historic perspective based on the European experience. Countries today tend to be more service orientated from the outset and skip a significant part of evolution.

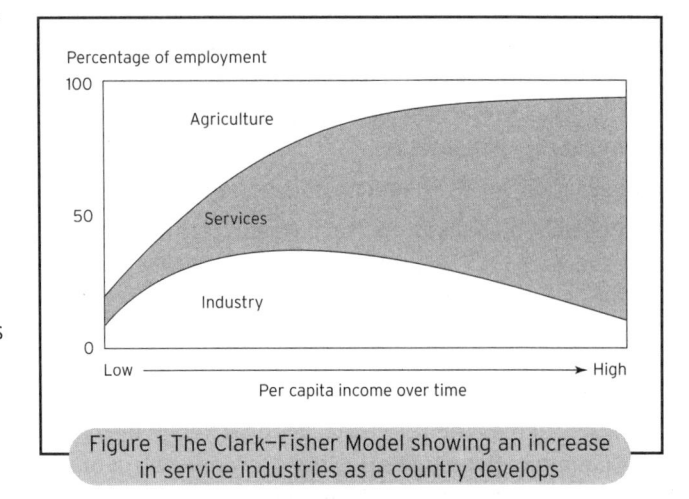

Figure 1 The Clark–Fisher Model showing an increase in service industries as a country develops

Service economies

The table below shows the differences between UK counties in the number employed by high technology and service industries.

The global patterns of service contribution to GDP (Figure 2) reflect the role of the state in the former socialist countries besides the range of service jobs available in MEDCs.

Deindustrialisation

There have been at least five important stages or waves of economic development, each one leaving an imprint on people and their economy. Each stage or **Kondratieff Wave** has resulted in the growth of industries and industrial regions.

- **K1** was the early industrial revolution.
- **K2** was the age of **steam power** that commenced in the 1840s. It gave rise to the railways and provided power for the mechanisation of textiles and the making of steel ships. It is associated with the coalfield industrial regions such as South Wales, the North East, the Ruhr in Germany, Nord–Pas de Calais in France, Sambre Meuse in Belgium.
- **K3** was the age of **electrical engineering** that commenced in the 1890s and enabled industry to be more market orientated often around the larger cities.
- **K4** was the introduction of mass production or **Fordism** in the 1930s. This wave is still present and was most manifest with the development of the large car assembly plants.

1.14 Changes in the location of economic activity

County	% employees	Number (000s)	% Growth 1991–2000
Berkshire	21.3	94.4	64.5
Cambridgeshire	15.3	51.6	28.9
Oxfordshire	15.2	48.0	82.5
Warwickshire	15.0	32.7	ND
Hertfordshire	14.7	72.9	ND
Buckinghamshire	14.0	61.8	26.9
Cheshire	13.9	61.8	ND
Wiltshire	13.9	38.6	40.6
Bedfordshire	13.5	30.0	ND
Surrey	13.1	73.9	24.2
ENGLAND	10.4	2.26 millions	3.8

Other high growth counties are East Sussex 18.1 per cent, Shropshire 15.1 per cent, Greater London 14.7 per cent and Nottinghamshire 13.1 per cent. Factors influencing this pattern: universities as sources of technology transfer companies, new labour and ideas; government research institutes nearby; multiplier effect; agglomeration; environment for business.

Table 1 Employees in high-tech manufacturing and services in the UK by county, 1991–2000

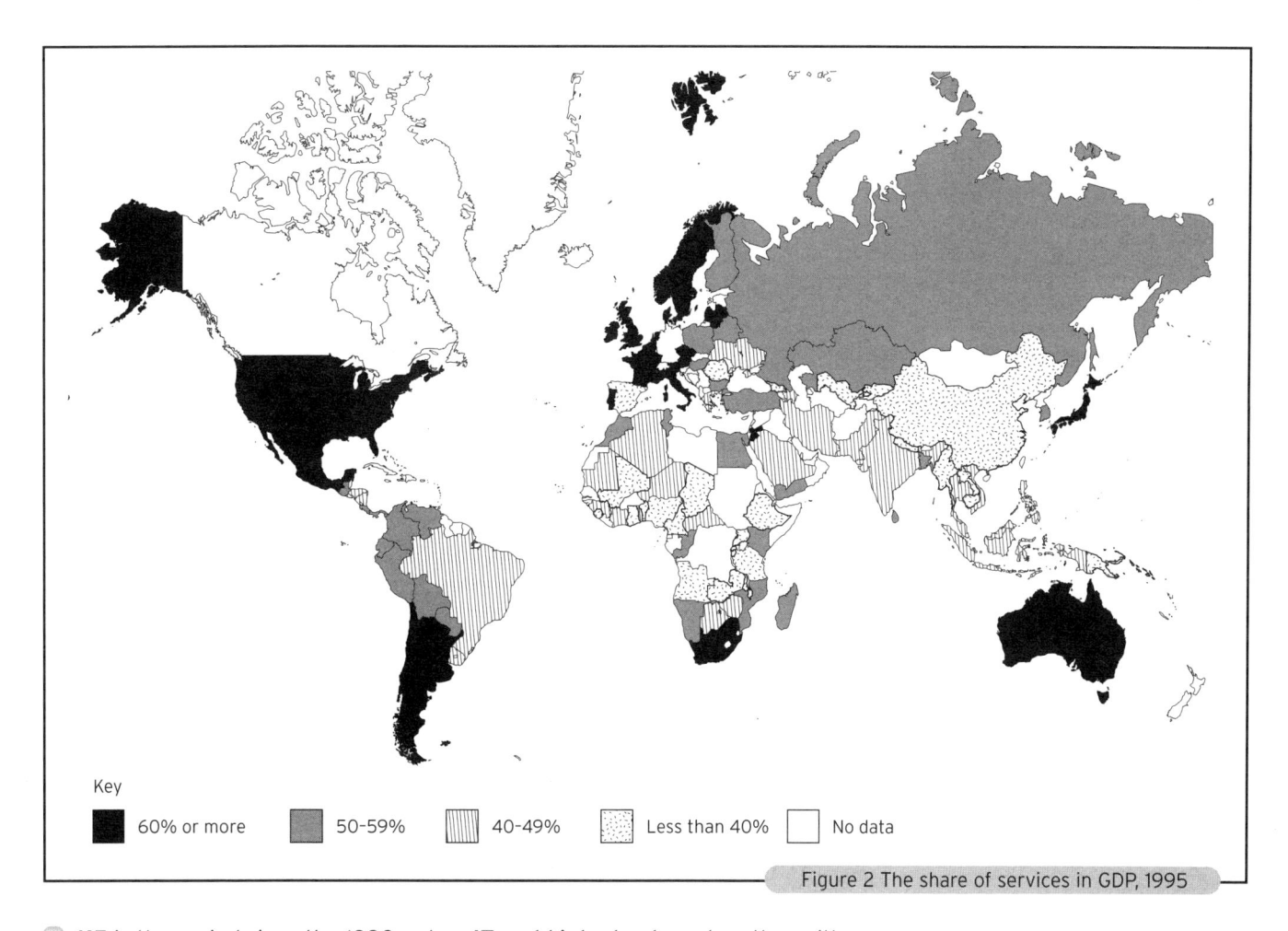

Figure 2 The share of services in GDP, 1995

Key: 60% or more ■ | 50–59% ▨ | 40–49% ▥ | Less than 40% ⋯ | No data □

- **K5** is the period since the 1990s when **IT and biotechnology**, together with satellites and digital networks, have brought about a new range of activities.

The activities that dominated in K2 and K3 have declined leading to **deindustrialisation** and have been replaced by newer activities. On the whole new activities prefer to go to new locations and, therefore, the old industrial buildings are abandoned because they are unsuitable for modern technologies. Governments can ameliorate the effects of decline by offering Regional Development Aid although this rarely replaces the jobs or retains the old skills. Therefore, the cycle of production rising, peaking and declining as new products are invented, **the product life cycle**, has a parallel set of changing locations and employment characteristics. Changing production cycles are just one component of deindustrialisation and the decline of regionally important industries. Other causes of deindustrialisation are:

- Competition from overseas, e.g. less costly ship building in Japan and Korea that destroyed much of ship building in the UK
- Products at end of their life cycle and not in demand
- Poor and outmoded Human Resource Management, e.g. job demarcation
- The need to rationalise production
- New technologies such as Just in Time removing the need for parts warehouses
- Removal of government support for old industries with subsidies
- Move of production to NICs and, more recently, to East Europe
- Products introduced more rapidly (Figure 3).

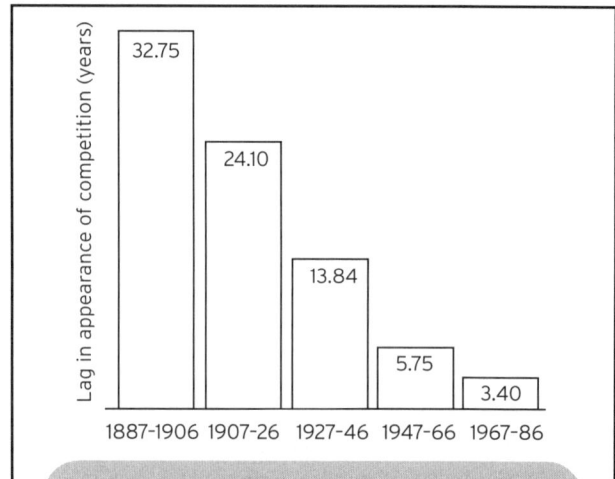

Figure 3 Bright ideas: interval between the introduction of an innovation and competitive entry

Global rank 2003 (Rank 2002)	Company	Country	Sector	Turnover $ millions
1 (2)	Microsoft	USA, Seattle	Software and computer services	28 365.0
2 (1)	General Electric	USA, Fairfield, CT	Range of industries	130 685.0
3 (3)	Exxon	USA, Houston	Oil and gas	204 506.0
4 (4)	Wal Mart Stores	USA, Arkansas	Retailers	244 524.0
5 (6)	Pfizer	USA, New York	Pharmaceuticals and biotechnology	32 373.0
6 (5)	Citigroup	USA, New York	Banking	ND
7 (9)	Johnson and Johnson	USA	Pharmaceuticals and biotechnology	32 298.0
8 (10)	Royal Dutch/Shell	UK/Netherlands London	Oil and gas	179 423.0
9 (8)	BP	UK, London	Oil and gas	178 721.0
10 (12)	IBM	USA, New York	Software and Computer Services	81 168.0
11 (11)	American International	USA, New York	Insurance	ND
12 (15)	Merck	USA, New Jersey	Pharmaceuticals and biotechnology	51 790.0
13 (17)	Vodafone	UK, Newbury	Telecommunications	35 818.7
14 (21)	Procter and Gamble	USA, Cincinnati	Personal care and household products	40 238.0
15 (7)	Intel	USA, Santa Clara	IT hardware	26 764.0
16 (13)	Glaxo Smith Kline	USA, North Carolina, Research Triangle Park	Pharmaceuticals and biotechnology	33 258.3
17 (22)	Novartis	Switzerland, Basel	Pharmaceuticals and biotechnology	23 606.5
18 (29)	Bank of America	USA, New York	Banking	ND
19 (14)	NTT Do Mo Co	Japan, Tokyo	Telecommunications	43 055.0
20 (16)	Coca Cola	USA, Atlanta	Beverages	19 564.0
23 (27)	HSBC	UK, London	Banking	ND
26 (28)	Toyota	Japan, Tokyo	Automobiles	125 765.3
27 (34)	Nestle	Switzerland, Zurich	Food producer and processor	64 937.4

Table 2 The Top 20 Companies ranked by their share value 2003

1.14 Changes in the location of economic activity

The age of the Trans-national Company (TNC)

Table 2 gives the ranking of the top companies in the world.

It is notable that most of the top companies are located in the USA. In fact 48 per cent of the global 500 are US based compared with 9.4 per cent in Japan, 4.4 per cent in Canada and 6.8 per cent in the UK. It is also notable that the big companies are not those that geographers normally study as examples of TNCs. Three companies are oil and gas producers, five are biotech companies and three are IT companies. The largest car producer is only twenty-sixth in the world. The USA has the largest company in every economic sector except telecommunications where Vodafone is the market leader.

The role of government in inward investment

Inward investment is sometimes called **Foreign Direct Investment (FDI)** which is where a company has a financial interest in an economic activity in a foreign country. Very often this is directed by governments who are attempting to attract investment capital to a country or a region. The table below shows some UK attempts to attract inward investment by means of regional selective assistance.

Where	Company	Date	Offered	Received	Jobs (remaining by 2002
Newport	LG Group	1997	£247m	ND	400
Mossend	Chungwa	1997	£100m	£15m	25
N Tyneside	Siemens	1996	£25m	£18m	1567
Bathgate	Motorola	1993	£57m	£45m	120

Table 3 Examples of companies supported by the UK government with regional selective assistance

These cases were all trumpeted as successes when set up but all have not been very successful. The LG chip plant was built but never opened and the TV tube plant shared with Phillips closed in 2003. Chungwa, a Taiwanese electronics firm was funded to make Cathode ray tubes and employ 3 300 but it closed in 2003. It never employed more than 1 200 people. Siemens never started production and paid the money back. Motorola also closed their plant making cell phones in 2001.

Benefits and costs of FDI for MEDCs
Benefits

- It has helped to regenerate areas that had deindustrialised and become run down, e.g. firms went to the Ruhr in Germany or the UK coalfields.
- The industries are normally ones that export and therefore benefit the national economy, e.g. Nissan supplies cars to Europe from Washington.
- New management skills are introduced through the new investments, e.g. Just in Time technologies.

Costs:

- The profits of FDI are returned to the country of ownership.
- It often creates a branch plant economy where firms can close the branches to maintain the core business at home (see LG above).
- Firms want to get into a trading bloc to gain advantages, e.g. Toyota in France and the UK.

Figure 4 shows the value of inward investment on countries and regions in Europe. The pattern is very volatile and these figures for 1998–9 were at a high point for FDI. Currency shifts, labour costs and political decisions (e.g. UK not joining the Euro zone) have altered the patterns since 1999.

Global challenge

Figure 4 Inward investment 1998–99

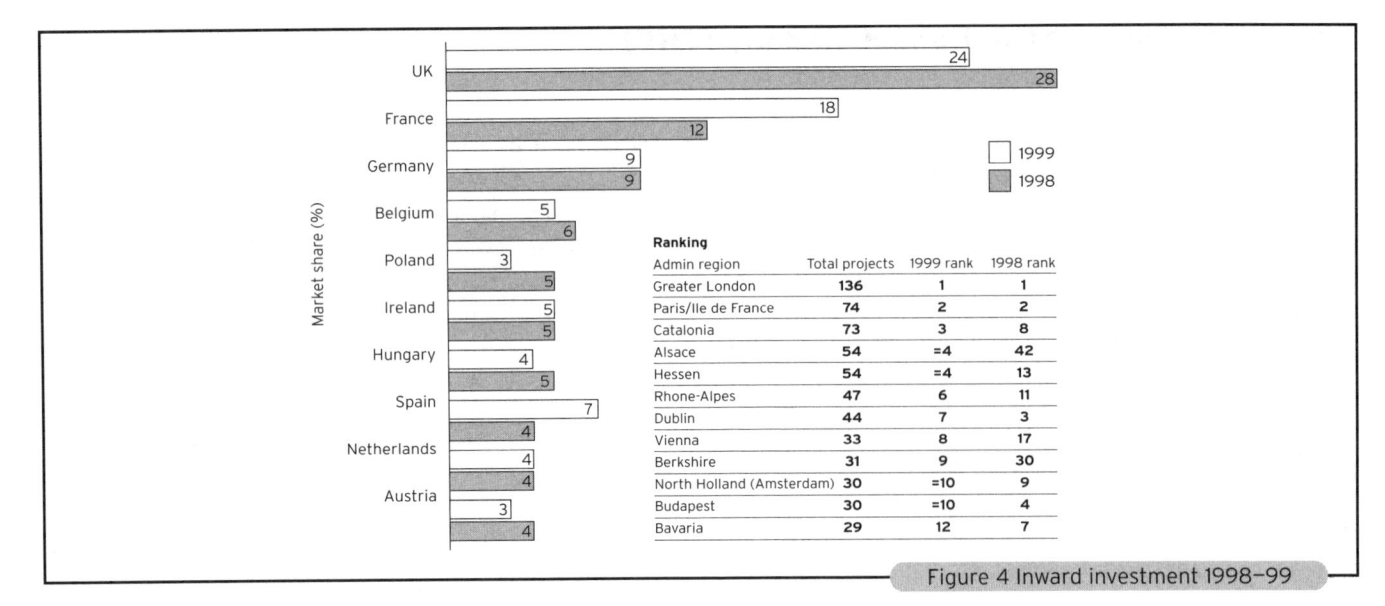

Market share (%) – 1999 and 1998:
- UK: 24 (1999), 28 (1998)
- France: 18 (1999), 12 (1998)
- Germany: 9 (1999), 9 (1998)
- Belgium: 5 (1999), 6 (1998)
- Poland: 3 (1999), 5 (1998)
- Ireland: 5 (1999), 5 (1998)
- Hungary: 4 (1999), 5 (1998)
- Spain: 7 (1999), 4 (1998)
- Netherlands: 4 (1999), 4 (1998)
- Austria: 3 (1999), 4 (1998)

Ranking

Admin region	Total projects	1999 rank	1998 rank
Greater London	136	1	1
Paris/Ile de France	74	2	2
Catalonia	73	3	8
Alsace	54	=4	42
Hessen	54	=4	13
Rhone-Alpes	47	6	11
Dublin	44	7	3
Vienna	33	8	17
Berkshire	31	9	30
North Holland (Amsterdam)	30	=10	9
Budapest	30	=10	4
Bavaria	29	12	7

The effects of global shift

Outsourcing

The Flight to India

Half of the world's top 500 companies now outsource IT or other business processes to India (estimated $24 billion benefit to the Indian economy by 2008).

Where in India?

Bangalore, Gurgaon near Delhi, Chennai, Hyderabad.

Why?
- Educated workforce: 1.29 million graduates a year
- Cost, e.g. sending a car hire call to Bangalore rather than a UK centre costs a total of 20p for the whole transaction
- Overall costs, up to 50 per cent cheaper – wage costs lower. An IT manager in USA costs $55 000 whereas in India $8 500 according to *Time* magazine.
- Multiplier effect of investments in IT in Bangalore draws in more similar companies
- Skills: IT very strong in education system.

Costs to India
- Westernisation and loss of cultural identity
- Monbiot talks of 'abandoning identity and slipping into someone else's'; dealing with abuse from angry customers
- Unsocial hours due to time differences
- Most employees leave after three years but without skills that can be transferred to other jobs in India
- Increases social divisions in India.

Benefits to India
- Young workers with a higher disposable income
- Gender apartheid being reduced
- Employment with higher starting salaries.

Costs to UK
- Job losses – 200 000 by 2005 (500 000 from USA).

Benefits to UK
- Improved profits of companies benefit shareholders such as pension funds.

Examples: HSBC (4000 in October 2003), BT, BA, Lloyds TSB, Prudential, Standard Chartered, Norwich Union, Bupa, Reuters, Abbey National, Powergen, Goldman Sachs.

1.14 Changes in the location of economic activity

It is not just to India as Figure 5 shows. In the Philippines billing for AOL, software production for Arthur Andersen, accounting for Alitalia and the Red Cross, engineering design for Mitsubishi, and on-line purchasing for Barnes and Noble have recently been set up. The companies have gone to the Philippines because there was a government drive to attract back-office operations. Companies gained tax breaks and specially designed back-office parks have been built for them in Manila and Cebu.

Many argue that economic globalisation is not matched by social globalisation and point to the opposition to labour migrants.

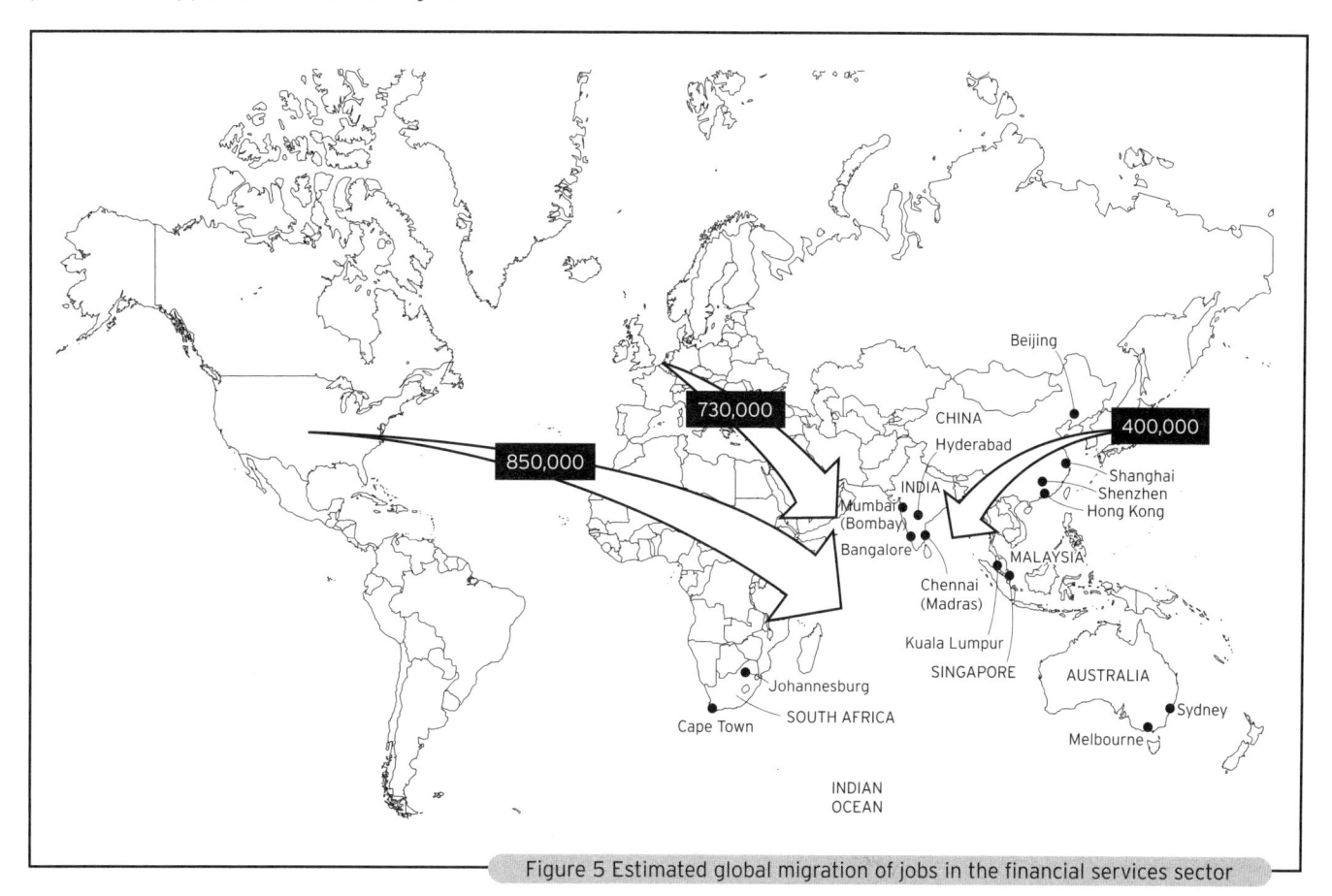

Figure 5 Estimated global migration of jobs in the financial services sector

QUICK CHECK

1. What effects might global shift have on the environment in two contrasting countries?
2. What challenges does global shift present to economies and peoples in contrasting countries?

1.15 The future of the global economy

Key question for this section:
- How might the global economy change in the future?

In this optional section you will be revising:
- Economic expansion
- Changes in the workplace
- Tertiarisation.

This option links back to your work in 1.13 and 1.14 on the changing nature of the world's economy.

Economic expansion

Trade is a sign of growing global economic integration in the eyes of the World Bank but of growing world divisions among the critics of current policies.

Where is economic expansion taking place?

1. In NICs since the 1980s due to state planning for inward and home-based investment in electronic technologies.
2. In RICs, e.g. China due to policy changes in the Chinese government that now permits inward investment by multinationals.
3. By **technological transfers,** e.g. from the USA to Mexico **maquiladora centres** and Japan to Malaysia (Proton cars from Mitsubishi).
4. By outsourcing of activities such as back-office routine work and call centres.
5. In MEDCs, through the changing nature of the economy and increasing number of jobs in IT and biotechnology (see 1.14, pp 38–63).

Changes in the workplace

In MEDCs the following changes are typical:

- More service-based jobs rather than production based. These range from new activities in finance to house sitting while on holiday. Figure 2 in 1.14 shows the global distribution of service industry GDP.
- More research and development work linked to companies and government. The M4 corridor is full of governmental research activities such as Aldermaston Atomic Energy Research and the Road Research Laboratories at Bracknell.
- Growth of the conference industry because more wealth is based on ideas and information. Conference centres are a part of most cities and meetings facilities are found in most hotels.
- Growth of exhibition centres and exhibition halls that double up for marketing, e.g. The Motor Show and pop concerts at the National Exhibition Centre (NEC), Birmingham. Setting up and moving such events is a growth activity.
- More consumerism: industries catering to lifestyles have become more common. Designer clothes shops, retail outlets, health clubs are all manifestations of lifestyle employment, e.g. Gunwharf Quays in Portsmouth has outlets for Polo Ralph Lauren, Levi, Ted Baker, Tommy Hilfiger, Gap, Paul Smith among others.
- More women participating in the labour force. This is traditional in subsistence economies, it is only Islamic countries that have a low percentage. Most women in MEDCs delay childbirth to pursue a career and then return to work after childbirth. New jobs in child care are growing to cater for children while both partners are at work.
- Computers have changed the nature of work and triggered the arrival of new types of work, e.g. working at home. Servicing the telecommunications industry has grown in parallel.
- Fewer over 55-year-olds in work because of good pension provision and savings. However, the downturn in stock market performance may reverse this trend.

1.15 The future of the global economy

- Decline in size of workforce, allied to the declining birth rate. Countries are now considering altering pension regulations to enable workers to remain in the workforce after the legal retirement age.
- Increase in part-time work, flexitime and job share appointments.
- Increased affluence causes demand for garden design and maintenance and interior design, which is often allied to lifestyle promotion in the media.
- New activities to cater for the changing population structure, e.g. budget airlines.
- Jobs in culture and education form the largest employment group in most of the developed world's cities.

Changes in work in LEDCs and LDCs include:

- Increase in the cultivation of cash crops and agribusinesses to satisfy global market demands which reduces the number of people subsistence farming.
- Exposure to unhealthy pesticides in new agribusinesses.
- Migration to main centres of employment in production line work in NICs, e.g. Batam Island, Indonesia where Singaporean companies employ cheap labour and workers live in barracks.
- Increase in sex tourism despite being condemned by many.
- Growth of cultural work related to tourism e.g. traditional dance groups.
- Growth of westernised work, e.g. Filipino bands playing in resort hotels in SE Asia.
- Low wage work in tourism, and the lack of work and underemployment in many LEDC cities.

> **Reminder**
>
> The key theme above is the **tertiarisation** of the work force in the latter stages of the Clark Fisher model (see 1.14, pp 58)

Geographical issues related to the workplace

Challenges in the workplace include:

- Automation and robotisation is a constant threat. Robots brought on redundancy at car plants in the last quarter of the twentieth century.
- Round-the-clock, 24/7, working has less respect for cultural norms such as holy days and the nature of family life.
- Working in the global economy when teleconferencing with Japan and the USA has to take account of time zones.
- The increase in movement between jobs and between careers. Many teachers arrive in mid life from careers in other service sector jobs.
- More migration to find work which can be within a country or between countries as economic migrants. Illegal migration and people trafficking is a further consequence of the perceived employment benefits.
- Possible ethical issues of working for companies whose products are created by new and challenging processes. Biotechnology jobs fall into this category especially for those connected with GM crops.
- The increase in demand in MEDCs for new support activities such as conferences, banking and investment opportunities, cabling and technology to support, e.g. broadband.
- To avoid the minimum wage.
- The issues of low wages and gender wage differentials.
- Increased use of contracted, limited-term labour rather than permanent staff.

> **Reminder**
>
> These issues will change and so it is worth keeping your eyes open for good examples of changing challenges in the workplace. Do consider the issues in relation to the levels of economic development.

Quick Check: Apply these changes to a country that you have studied.

1.16 Addressing the development gap

Key question for this section:

- How might the development gap be addressed?

You will be revising:

- The trade gap and its effects on LEDCs
- How to close the gap and organisations assisting development
- The role of NGOs.

This option links to the key ideas in 1.13 concerning the development gap and globalisation. It presupposes that your knowledge of 1.13 is good.

The trade gap and its effects on LEDCs

These are some examples of the extreme differences between the richest and poorest countries.

- In 2000 the annual subsidy for a dairy cow in the EU was $913. The annual aid to a person in Sub-Saharan Africa from the EU was $8. That person's average income was $490. The EU provides double the development aid of USA.
- Real incomes of the 1 billion living in high income countries are 75 times those of the 1.2 billion living in poverty on $1 per day. The number living in poverty has declined between 1990 and 2001 by 210 million people, which is mostly due to the growth of the economy in 30, mainly Asian, countries and especially China.
- 54 countries with 12 per cent of global population had declining economies in 2000.
- The profits of the top 10 companies are equal to the GDP of 29 African countries.
- The wealth of the world's 11 richest people is more than the combined GDP of 49 LEDCs.

> **Reminder**
>
> Check out the role of the organisations listed in Table 2, 1.14, with regard to addressing the gap. Their websites are easy to find and are very comprehensive.

UNDP Millennium Development goals and targets

The UN Development Program decided on 18 goals in total but Goal 8 is seen as the most important because it complements the first seven:

> **Goal:**
> Goal 8 calls for an open, rule-based trading and financial system, more generous aid to countries committed to poverty reduction, and relief for the debt problems of developing countries. It draws attention to the problems of the least developed countries and of landlocked countries and small island developing states, which have greater difficulty competing in the global economy. It also calls for cooperation with the private sector to address youth unemployment, ensure access to affordable, essential drugs, and make available the benefits of new technologies.

> **Targets:**
> **Target 12** Develop further an open, rule-based, predictable, non-discriminatory trading and financial system. It includes a commitment to good governance, development, and poverty reduction – both nationally and internationally.
>
> **Target 13** Address the special needs of the least developed countries. It includes tariff and quota-free access for least-developed countries' exports; an enhanced program of debt relief for HIPCs (Heavily Indebted Poor Countries) and cancellation of official bilateral debt; and more generous ODA (Overseas Development Aid) for countries committed to poverty reduction.
>
> **Target 14** Address the special needs of landlocked countries and Small Island Developing States (through the Program of Action for the Sustainable Development of Small Island Developing States of the twenty-second special session of the General Assembly).

1.16 Addressing the development gap

Target 15 Deal comprehensively with the debt problems of developing countries through national and international measures in order to make debt sustainable in the long term.

Target 16 In cooperation with developing countries, develop and implement strategies for decent and productive work for youth.

Target 17 In cooperation with pharmaceutical companies, provide access to affordable essential drugs in developing countries.

Target 18 In cooperation with the private sector, make available the benefits of new technologies, especially information and communications.

Declining GDP

The ease of an economy growing rapidly until the bubble bursts and slipping back is ever present. Agricultural subsidies in MEDCs are blamed by the UN for the continuing economic crisis in Africa. As a consequence only five countries in Africa had a sufficiently high growth of GDP in 2002 (in excess of 7 per cent) to meet the Millenium goals. Five countries had declining GDP. The worst cases of declining GDP are where civil conflict and war are rife (Zimbabwe and Liberia). Where flooding and drought threaten crops GDP growth was also low, e.g. Ethiopia as Figure 1 shows.

Closing the gap

Possible methods of alleviating poverty include:

- The **Tobin Tax**, which is a tax on international flows of money that could be used to fund development. It has not been implemented but it would provide funds from the global economy for development.
- **Jubilee Research**, the successor to **Jubilee 2000**, which campaigns for the cancellation of international debt advocating 100 per cent cancellation of the unpayable and uncollectable debts of developing countries.

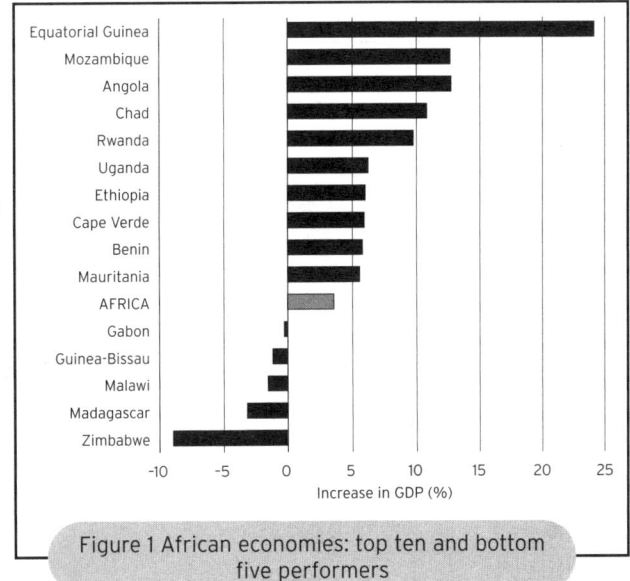

Figure 1 African economies: top ten and bottom five performers

It also aims to make international financial activities of sovereign governments and multilateral institutions more transparent and accountable to citizens. It highlights environmentally sustainable policies for financing development and paying the ecological debt of the north to the south.

Top down or bottom up?

1 Top down
- Involves strong central government direction – planned economies were characteristic of the former Soviet Bloc but it is also a characteristic of NICs such as Malaysia.
- Can involve large sums of inward investment, e.g. US investment in Akosombo because linked to aluminium production; the Pergau dam project.
- The theory is that the benefits accruing will trickle down and lead to local and regional economic development.

2 Bottom up
Often generated by local initiatives
- Some schemes might be encouraged by aid agencies, e.g. rope wells
- It depends on the benefits being easily apparent so that the idea is rolled out to other regions.
- Can often follow humanitarian disasters when local capacity is most receptive.

3 Localisation
Proposed by WTO – 'everything that can be produced locally should be produced locally' – intended to help by making people increasingly self-reliant. However, it is impractical because it assumes all small countries have the raw materials and it would increase costs for small countries with no large market.

The role of NGOs in approaching poverty

NGOs have done much to raise awareness. Examples include:
- Friends of the Earth who have attacked the 'potent cocktail' of increasing corporate power and weakening governmental power.
- The **Girona Declaration** signed by 40 campaigning NGOs in 2003 criticised companies for trying to '**greenwash**' decisions, i.e. professing environmental good when there is no substance.
- Oxfam's petition 'to make trade fair'.

NGOs can only be successful if they remain outside of government influence.

NGOs have also demonstrated how thinking small and using appropriate technology are often the most beneficial methods of dealing with poverty. Schemes such as the construction of rope wells often bring more benefits to local people than other more grandiose schemes.

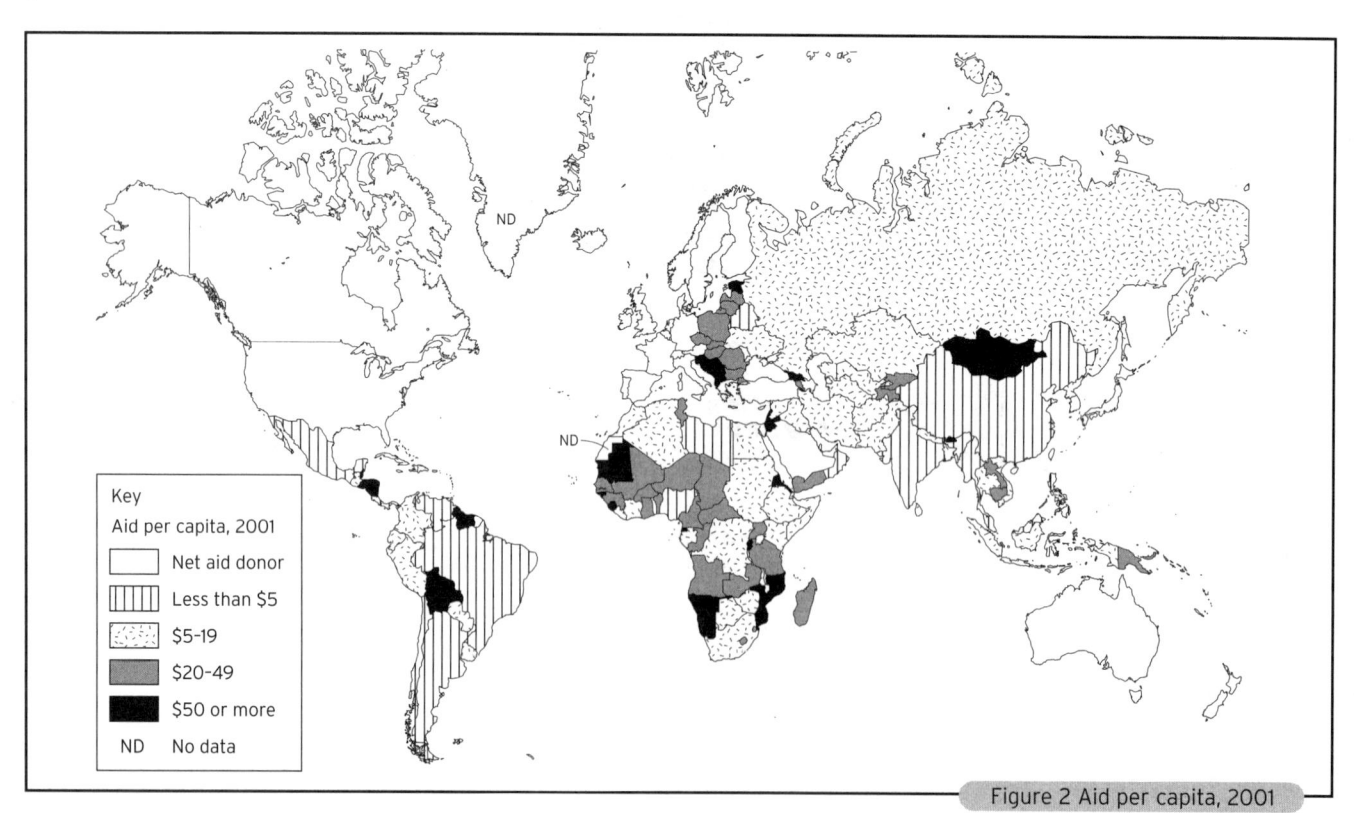

Figure 2 Aid per capita, 2001

1. Compare the relative merits of top-down and bottom-up approaches to development.
2. Comment on the data in the map in Figure 2 in terms of the prime destinations. Are these destinations the right ones? Why might aid be misplaced in terms of real need?
3. Discuss the efforts of an NGO or a charity in the field of development. Do they help to close the development gap?

1.17 Sustainable development

Key question for this section:
- Can sustainable development be achieved?

You will be revising:
- Global tensions
- Green growth and consumerism
- Economic, environmental, social and demographic sustainability.

This option links back to section 1.14. It is a key option in terms of providing help with the cross-unit questions.

Global tensions

Failed summits, initiatives and targets

Tensions were expressed at the Johannesburg Earth Summit 2002 where TNCs and local communities felt that they had done more than governments. Not all companies have been taking account of social and environmental factors according to UNEP. Economic growth often overtakes the efforts of businesses, e.g. in the field of global warming.

There is talk of a **sustainability gap** brought about by the effects of increasing affluence and the over-exploitation of ecosystems. The worst effects are felt by those who are least prepared and are the poorest. When the least prepared countries destabilise they can become a security risk and threaten the rich. Many look at the legacy of Rio 1992 and point to the slow rate of progress. Despite the Convention on Biodiversity 1993 species loss is accelerating, and countries have been unable to agree on a convention on forest exploitation.

The Global Reporting Initiative 2002 is an attempt to formalise reporting on non-financial performance. The guidelines suggest 110 measures of environmental performance relating to sustainable performance. 120 companies have adopted the idea, e.g. ICI and Isuzu Motors. But, because it is voluntary, many companies will continue to ignore the initiative and to act in an unsustainable way.

Resource provision versus resource consumption

Famine in Southern Africa affects 13 million people in 7 countries who do not have access to food. Maize harvests declined by 77 per cent in Zimbabwe and 42 per cent in Zambia in 2001–2. The Millennium Target was to halve malnutrition by 2015 for 400 million people. The UN Food Summit in June 2002 declared that on current trends 600 million would still be malnourished in 2015. Land available for food production, especially in LEDCs, will not increase especially in North Africa and South Asia. Water is an even scarcer resource. Already agriculture takes 90 per cent of the water in LEDCs. At least 20 per cent more water is needed to feed the expected population increases until 2020. Much of the land where food can be grown is being used for cash crops for export.

Economic and environmental sustainability

There is a conflict between these two aspects of sustainability. A high quality of life is normally found where there is a high consumption of resources. Unless a country is able to exploit and replace its own resources, it will depend on resources bought from elsewhere. Most MEDCs are more destructive of the environment in LEDCs than the LEDCs themselves.

Green growth

This economic and social development cannot be sustained without any markedly adverse effect on the environment. Various approaches to green growth have been adopted.

> **Reminder**
>
> **Sustainability gap**
> The gap between the poor and rich where the poor are the worst affected by the excesses of the rich.
>
> **Sustainability**
> See the definition of sustainability on pages 36, 73 and 141.
>
> **Green growth**
> The reliance on development that is environmentally sustainable.
>
> **Green consumerism**
> Using the consumer to force more environmentally and economically sustainable policies on companies.

Corporate Social Responsibility was introduced by the EU in 2001 to enable companies to contribute to a better society and a cleaner environment. It is an approach to economic and social sustainability. Suggested ways of achieving the goals are:

- **Eco-efficiency** – improve the way in which the use of resources can reduce environmental damage and at the same time reduce costs.
- **Eco-audits** – applying non-financial criteria to investment decisions,
- **Ethical audit** – applying non-financial ethical criteria to investment, the shares held by investment companies and the companies with which one trades.
- **Fair Trade** – trading that promotes sustainable development for those that conventional lowest price trading excludes and disadvantages.
- **The triple bottom line** – companies should be judged not just on economic prosperity, but their contribution to environmental quality and social well-being.

Green consumerism

This might have the power to force governments and producers to be more proactive. Buy eco-friendly, e.g. light bulbs that have longer lives and do not have built-in obsolescence. Buy organic vegetables. The Brent Spar incident showed consumer power in 1995 when consumers boycotted Shell because it wanted to sink the oil platform. Shell relented and broke it up in Norway. In USA, Green Gauge Report found that many said they were green but only 29 per cent actually made green purchases.

Alternative agricultures

Some examples include:

- **Crop-based proteins,** i.e. growing crop-based proteins such as soya rather than feeding crops to animals to produce meat and milk.
- **Organic farming** with an absence of inorganic fertilisers and pesticides and the use of organic manures. It has low inputs but higher prices which may not be affordable. It is sometimes called **permaculture.**
- **Irrigation** was once the solution. However, 60 per cent of irrigation water is lost to evaporation and 20–30 per cent of irrigated land in LEDCs has been damaged by water-logging and increased salinity.
- **The Green Revolution** did increase yields. New initiatives such as **GM crops** are all privately led, often by TNCs, e.g. Monsanto.

Ecotourism

Tourism accounts for 11 per cent of global GNP and employs an estimated 200 million people. It transports 700 million travellers each year. It is the principle export of 49 countries and the largest income earner in 37 countries. No wonder that its sustainability is a necessity when the scale of tourism suggests that it threatens the very environments that people wish to visit. It is doubly necessary for countries that depend on tourism because they depend on the environment to attract visitors.

'Responsible travel to natural areas that conserves the environment and sustains the well-being of local people' is the definition offered by the International Ecotourism Society.

Ecotourism projects tend to be small-scale and only appeal to a minority of tourists because the costs are higher. Mass-tourism is still not eco-friendly because of the sheer scale of transport, water use, food demands that are made in the main resorts.

> **Quick Check:** Can the various types of sustainability work in the least developed countries? To what extent should MEDCs be prepared to help the LEDCs to create a more sustainable future for all?

Reminder

This is the key issue. If population continues to increase in the LEDCs and LDCs then the prime aim of sustainability, improved quality of life, might not be attainable. Globalisation both assists populations to improve their quality of life and also places demands on resources that impair the attainment of a better quality of life. Therefore, many are of the opinion that the unevenly distributed, rising global population is the most urgent Global Challenge.

1.18 Examination questions

Sections A and B, Paper 6474

All questions in Sections A and B have 25 marks. You must answer one from each Section. The questions are generally in two parts and the marks are a guide to the proportion of 35 minutes that you spend on each part. All questions contain stimulus material in the form of maps, diagrams, graphs, tables and photographs. Look at the resources carefully because they should trigger the response but not provide the whole answer. Very often the questions will use a cartoon to stimulate your ideas. The cartoons do not give you the answer but they should enable you to focus on the issue.

Figure 1 Outsourcing

1 Figure 1 shows a cartoon commenting on outsourcing.

(a) With the aid of examples from a range of economic activities explain what is meant by outsourcing. (10 marks)

(b) Examine the impact of outsourcing on countries at different stages of development. (15 marks)

Global challenge

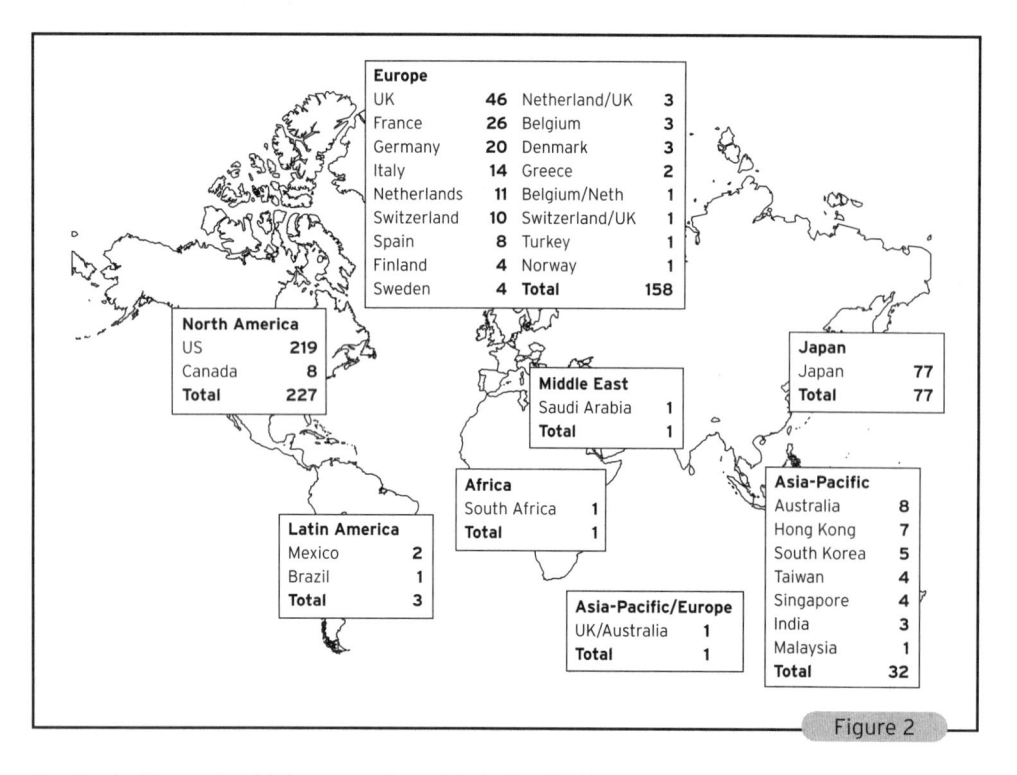

Figure 2

Europe
UK	46	Netherland/UK	3
France	26	Belgium	3
Germany	20	Denmark	3
Italy	14	Greece	2
Netherlands	11	Belgium/Neth	1
Switzerland	10	Switzerland/UK	1
Spain	8	Turkey	1
Finland	4	Norway	1
Sweden	4	**Total**	**158**

North America
US	219
Canada	8
Total	**227**

Middle East
Saudi Arabia	1
Total	**1**

Japan
Japan	77
Total	**77**

Latin America
Mexico	2
Brazil	1
Total	**3**

Africa
South Africa	1
Total	**1**

Asia-Pacific/Europe
UK/Australia	1
Total	**1**

Asia-Pacific
Australia	8
Hong Kong	7
South Korea	5
Taiwan	4
Singapore	4
India	3
Malaysia	1
Total	**32**

2 Study Figure 2 which shows the global distribution of the headquarters of the world's top 500 companies in 2002.
(a) Explain the pattern shown on the map. (10 marks)
(b) With reference to either ONE trans-national manufacturing or ONE trans-national service company, outline and explain the global distribution of its activities (15 marks)

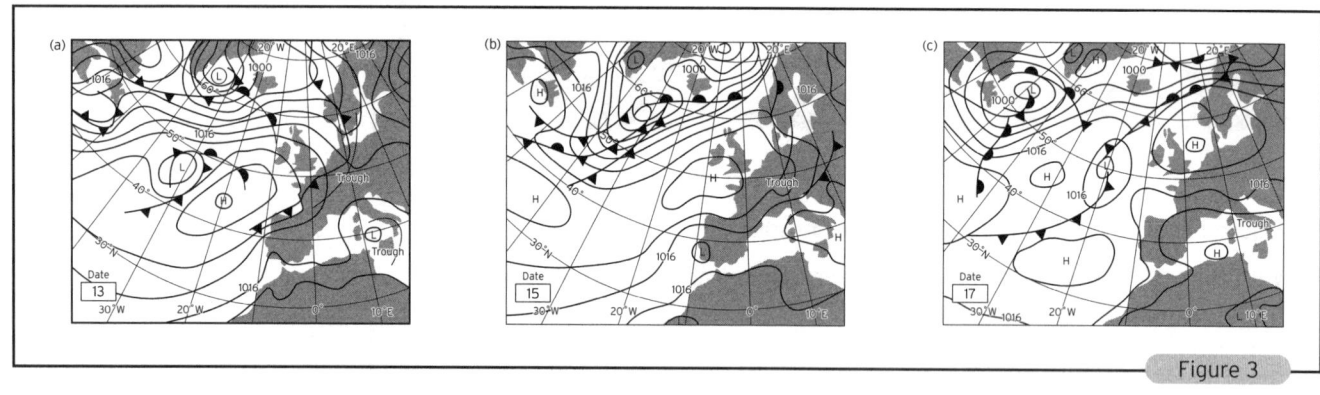

Figure 3

Question 3 relates to Section A on the examination paper. It is an example of a resource being a source of information and a stimulus for your response.

3 Study Figure 3 which shows the progress of a blocking high pressure system over the UK in August 1995.
(a) Outline the roles of the jet stream and air masses on the weather experienced during the slow north-eastwards passage of the blocking high over the UK. (15 marks)
(b) Describe and explain the differences between the effects of blocking highs in summer and winter on NW Europe. (10 marks)

1.19 Dealing with the cross-unit questions

Section C, Paper 6474

What are the key themes of Global Challenge?

1 Sustainability

It is defined by the World Commission on Environment and Development as 'the process of change in which the exploitation of resources, the direction of investments, the orientation of technological development and institutional change are all in harmony and enhance both current and future potential to meet human needs and aspirations'.

Development must meet the needs of people without depleting resources or damaging the systems which produce those resources be they environmental, or ecological or economic.

2 Conservation versus Development

This theme can be approached in several ways that embrace all global challenges.
- Conservation can be of ecosystems and biological reserves, or conservation of resources, or conservation of a variety of ways of life, or conservation of the balance between population and resources.
- Development may be economic, bridging the Development Gap, improving living standards, health care, social welfare and life expectancy.
- The conflicts are between those of economic advancement and the resources needed to maintain advancement.

3 Environmental degradation and destruction

This is the product of climate change, hazard events, exploitation of resources, the activities of governments and companies. Individual actions may also produce the same effects on a small scale and only become significant when they are combined.

4 Poverty

The poverty of environments may result from many of the activities described above. Climate change may result in poverty. Economic Development, or lack of it, may cause poverty because some countries do not have the resources, including human resources, to develop at the speed of the wealthy. Even within the most advanced societies poverty exists and the gap between rich and poor in spatial and social terms is widening.

5 Globalisation

It has an impact on environments. It is simplifying ecosystems, causing global warming, altering the geography of production and consumption and impacting on economies and people.

> **Reminder**
>
> Make sure you have a profile of a LDC that will illustrate the factors causing environmental, economic and social poverty.

QUICK CHECK

1. How far does the definition in the paragraph 'Sustainability' above improve on that of Bruntland?
2. Is it possible to have sustainability when economic advances depend on environmental exploitation?
3. Give a local example of the conflict between conservation and development?
4. Provide examples of environmental degradation at a variety of scales from the local to regional?
5. Give some short examples of globalisation's impact including some positive effects.

Global challenge

Section C questions

All Section C questions carry 30 marks. You should therefore allocate 45 minutes to answering your chosen question. A rough guide to time allocation is to allow 15 minutes for every 10 marks. The extra 5 marks from Section C (or at least some of them!) could get you a higher grade. Some good students actually answer Section C first so that they have it completed in the relevant time. There is no rule about the order of answers on this paper.

The question below is based on one set in 2003. It overlaps population and economy.

1 a) Describe and suggest reasons for the relationships between the data shown in the table. (12 marks)

Country	Per capita GNP$	Infant mortality rate	Population growth rate	Total fertility rate
Angola	410	292	3.3	6.7
Malawi	170	219	2.5	6.7
Somalia	120	211	3.9	7
Nigeria	260	191	2.8	6
Gambia	320	213	2.3	5.2
S Korea	9 700	9	0.9	1.8
UK	18 700	7	0.1	1.7
Singapore	26 730	6	1.5	1.8
Germany	27 250	7	0.3	1.3
Switzerland	40 630	7	0.7	1.5

NB Total fertility rate is the average number of children that would be born if all women in the country lived through to the end of their child-bearing years, normally accepted as 15–45 years old.

Table 1

Part (b) asks you to add consideration of weather, climate and ecosystems and to evaluate them as potential factors.

b) To what extent can environmental conditions explain the figures shown in the table? (18 marks)

The overarching themes
The poverty cluster bomb (figure 1) is an example of an overarching theme.

2 (a) With the aid of examples, discuss how the bomblets interact with one another to cause poverty. (15 marks)

You will need to look at several pairings in order to produce a convincing answer.

b Select any two bomblets and suggest how these might be tackled to reduce poverty. (15 marks)

Choice of an issue
The question above is one example. Very often the format will be like this part b of the question:

3 b 'Sustainability is the world's greatest challenge.' Discuss how this challenge may be met in the management of ONE of the following: the atmosphere, the biosphere, population, the economy. (15 marks)

Located issue in a country or region of your choice
The question below is the second part and enables you to select your own illustration:

4 b Discuss the effects that rising population may have on the economy and environment of coastal areas. (18 marks)

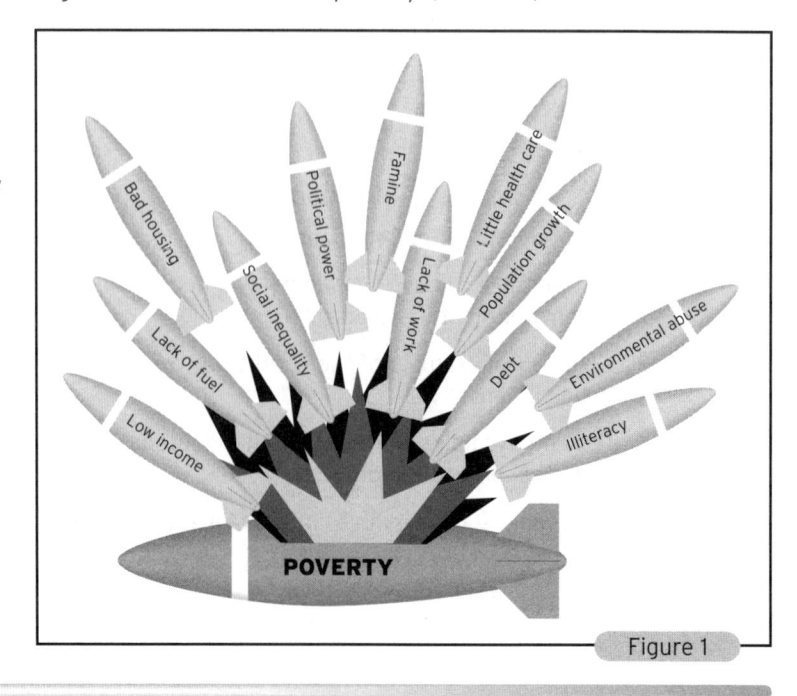

Figure 1

PART 2: Researching Global Futures: Managing natural environments

General principles and strategies

Part 2 covers the first paper of Unit 5 (6475/1): Managing natural environments. You will be expected to choose one of the following options:

Environments and resources (2.1, pages 78–80)
Living with hazardous environments (2.2, pages 81–93)
The pollution of natural environments (2.3, pages 94–108)
Wilderness environments (2.4, pages 109–118)

There are four generalisations for each option and, in the case of 2.2, 2.3, and 2.4, they are covered by four sections (one for each generalisation) in each (e.g. 2.2a, b, c and d).

Paper 1 (6475/1) is based on your research over a period of time – normally up to two months. Shortly before the examination, you will be told which generalisation will be the focus of TWO essay titles on your option. YOU ONLY HAVE TO ANSWER ONE OF THE QUESTIONS. You can focus your revision in the period after the generalisation has been issued. Don't spend too much time on the paper because it is only worth 7.5 per cent of the A Level mark. But do spend enough time. The mark is combined with that for Paper 2 (6475/2).

Selecting the option

Students may study any of the options. It may be unwise to study an option on your own; it is usually a good idea to have someone to bounce your ideas off and to shoulder some of the research burdens.

Getting started

1 Read the relevant textbooks to give you a feel for the topic. In particular focus on the principles in the generalisations when you take notes. The principles will be outlined in the sections on each generalisation.
2 Make use of the case studies and generalisation tables at the start of each option to help you organise your notes.
3 Read any articles from journals such as *Geography Review* and *Geographical* that refer to the topic that you are studying. Your college or school probably has these in resource files for you to consult. There might be articles in other magazines such as *New Scientist*, *Geo-news review*, *National Geographic*, and *The Ecologist*.
4 Start a media file. Assemble in this file articles from the quality press that refer to your option i.e. *The Times, The Guardian, The Observer, The Independent, The Sunday Times* and *The Daily Telegraph.* Do pick up on hazards and pollution events in particular. Gather travel brochures from the more specialist tour operators such as Turquoise, The Field Studies Council, Page and Moy, Cox and King, and Kuoni who all arrange holidays in wildernesses and use eco-tourist resorts. Sometimes it will pay to see if you can get the press for the country where an event is happening. *America Today's* weather pages are very good when hurricanes or tornadoes are around.

Assembling case studies

- Make sure that you have a range of case studies that illustrate your breadth of research. The same case studies will not fit all questions. For instance do not rely on just tectonic hazards, or just oil spills or just Antarctica because you will lose marks for poor research.
- Make sure that examples are appropriate for the points that you want to make.
- You will score in the research section if some examples are not those in every textbook.

Once the generalisations are released

- Review your notes on both the Foundation section and the generalisation to be assessed to see if there are gaps in your knowledge and understanding.
- Check that your examples are 'fit for purpose'.
- See if there are any sketch maps or diagrams that might enable you to illustrate your answer.
- Make sure that you can define the terms that are needed in an answer to the generalisation.
- Practise plans for questions on the generalisation in past papers. However, do not learn these essays because it is unlikely that they will appear in exactly the same format.
- Try writing answers in the time so that you are used to the timescale.

See the revision checklist below – confirm that you have completed the tasks.

In the examination

1. Read the question, highlighting the subject and command words.
2. Put your ideas onto the answer book and then order them into a plan.
3. Write an introduction to the topic, which defines the terms in the question and indicates the main thrust of the answer. State which examples you are using and why.
4. Following the plan write your answer using examples to support the points that you are making. Do not write a set of exemplar case studies and try to tie them to the question at the end. This scores lower marks for understanding.
5. Write a conclusion that sums up your essay. It should show that you have answered the question.
6. If you are running out of time, leave a gap and write a conclusion. Remember that the conclusion gains up to 10 marks. You will get more marks if you conclude rather than continue with the core of the essay. In the case of incomplete essays, the markers are told to look at the plan.
7. Try to show where you obtained material from – the name of an author, the website domain, not the details. This will indicate that you have researched your topic.
8. There is a world map provided. Use this map if you have to show a distribution, e.g. hurricane paths or plate boundaries. DO NOT use it to show the locations of your case studies. Locations are a hostage to fortune especially if you locate a place on the wrong continent!

Revision checklist

Tick the boxes when you are certain that you have all that you need. The second to last column is filled in last as a final check. Do not delude yourself; be honest.

You will need to refer to 2.1 to 2.4 (pages 78–118) to know what the generalisations state for each option.

Checks Global Futures Unit 5, Paper 6475/1	Have organised notes	Have examples – some of my own	Read relevant textbooks	Know terms and can define them	Can draw sketch maps	Have diagrams and models	Completed my primary study	Can use for 6474
Foundations								
Generalisation 1								
Generalisation 2								
Generalisation 3								
Generalisation 4								
When generalisation for the exam released	Have organised foundation and notes for generalisation	Made certain that my examples are best for this generalisation	Textbooks have been reread	Have created my dictionary of relevant definitions	Have quick maps memorised	Have basic diagrams memorised	Completed revision for the generalisation	Not relevant when generalisations released

2.1 The environment and its resources

Key questions for this option:

- What are resources?
- How are decisions made to exploit them?

You will be revising:

- Resources and development
- Strategies and management.

This option links very closely to the Global Challenge paper (6474). Your teacher can expand on the key themes. You must have studied at least ONE mineral resource and ONE energy resource.

Resources and development

The starting point in this option is to know your key definitions and to be able to give examples of them.

Reserve	The amount of a resource available under current technological and economic conditions.
Actual resource	The resource currently being exploited.
Local or point resource	A resource confined to a small area or region such as Zambian copper or the Isle of Purbeck oil wells.
Ubiquitous resources	Those that are widespread such as sands and gravels and various types of building stone.
Sustainable resource	One that can be naturally replaced as fast as it is used. Timber extraction in parts of Scandinavia aims to be sustainable.
Renewable resources	Those that are continually available for use. They are subdivided into **stock resources,** short term resources such as potatoes, and **flow resources,** those that can be depleted and increased by human activity.
Recyclable resources	Those capable of reuse such as tin, copper, lead.
Finite resources	Those whose reserves are known and may run out if exploitation continues at the current rate.
Infinite resources	Those which will not run out because they are renewed or replenished.
Carrying capacity	The maximum number of users that are able to be supported by a resource given the current technology.
Product Life Cycle	The stages in the growth and decline of use of a resource or product.
Geological Resource	A resource that occurs naturally in the Earth's crust. It can be **localised** or **ubiquitous.**
Political resource	A resource which gives a country power over others e.g oil.

There are four generalisations for this option and the main points that you should make in your essays are outlined below.

Generalisation 1: How do resources affect economic development? What are the consequences of this?

- Resource distributions are not even. Resource-rich regions have developed as foci of industrialisation, e.g. coalfield industrial areas in nineteenth century and market-orientated industries in twentieth century. You should have studied the growth of one region. Tin was one factor that encouraged Malaysian economic development although oil really aided take off to NIC status.

2.1 The environment and its resources

Figure 1 Product life cycle

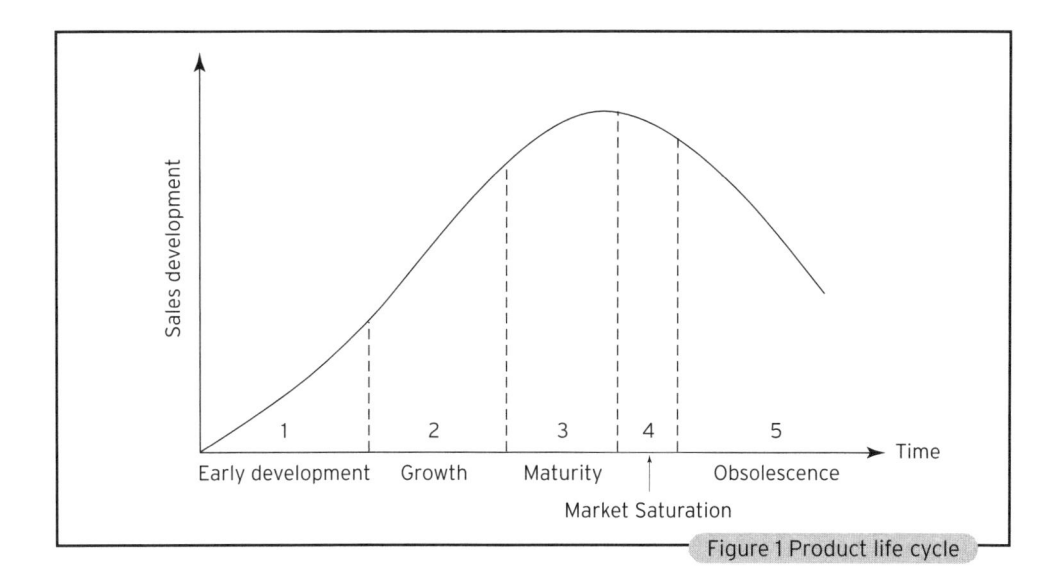

- The pattern of global energy production and consumption is not even. The consequences are regions of wealthy producers and regions of wealthy consumers and regions of want and poverty. It is similar for minerals although the effects on levels of development are less marked.
- What are the effects of resource exploitation on countries at various stages of development? What happens in terms of controls over mining when it is important to a country?

Generalisation 2: What are the environmental impacts of resource exploitation? Can these be successfully managed?

- If you live near a current or former area of exploitation it would be wise to tabulate the costs and benefits of exploitation on the area around the resource. Bedfordshire brick pits, Chichester gravel pits, Derbyshire and Somerset stone quarries and Cornish tin mines and china clay pits are all good case studies. Open-cast mining such as for brown coal in the brown coal triangle between Aachen, Koln and Munchen Gladbach, Germany would make a good regional case study.
- There does appear to have been a sequence of energy exploitation, coal, oil and nuclear in some kind of cycle. Why have countries moved through this cycle? Is it the exhaustion of reserves, pressure from the electricity companies, government policy that has caused countries to alter their energy use?
- Nuclear energy has always been controversial. Why do countries like France depend on it for 80 per cent of their electricity? What are the current issues in decommissioning nuclear power stations? Is it possible to manage such a risky business? How is nuclear waste managed?
- Sometimes impacts are negative such as Shell in the Ogoni region of Nigeria, and Alaska.
- Resource exploitation especially in countries without strong environmental protection legislation does have negative effects such as dredging for tin altering river flows, loss of habitats and species diversity, toxic waters from the tailings (waste dumps).
- Impacts such as increasing levels of carbon dioxide and acid rain.

Generalisation 3: What are the questions for the future of resource use and management? Can resources be effectively managed?

For this generalisation it is possible to specialise on either a mineral, e.g. tin or copper, or an energy resource, e.g. coal.

- Resources are being used unevenly across the world. The efficiency of their use varies according to the pressures on the resource, the cost of the resource and the ability of a country to optimise its use of a resource through technological advances and recycling. LEDCs are less able to benefit from technological advances in resource use. Cars in LEDCs tend to be more polluting and use more

energy than some of the energy efficient models being developed in Japan such as those for dual fuel use.
- How are reserves reassessed? In the North Sea, old oil fields are being revisited because the technology to extract more oil from what were unextractable resources has been developed. This could be seen as governmental pressure to maintain indigenous supplies rather than rely on imports or company innovations to maintain fields at locations in 'friendly' countries.
- Waste products are an increasingly important resource. Small economies have grown in LEDCs exploiting waste tips out of sheer necessity to have some income. Waste processing such as ship breaking in India, Pakistan and even Teesside is growing and causing controversy. Waste recycling is increasing because of the lack of landfill and international legislation such as compliance with Agenda 21.
- Reuse of materials that have been discarded such as cardboard, rubber tyres for a whole series of uses from shoes to road surfaces is growing. Does it save resources?

Generalisation 4: What are the resource use strategies for the future? What are the consequences of these strategies?

- Conservation of resources is one strategy for the future. Does conservation act only in the interests of MEDCs hampering LEDC development?
- Exploitation and technological advancement is the solution to resource depletion, an argument used by the Bush administration in the USA. It links to Boserup's 'necessity is the mother of invention' stance with regard to population.
- What are the social and economic consequences of conservation and exploitation?
- The importance of new discoveries, e.g. oil in Chad and its impact on the country and its exploiters. Can we continue to develop technologies to exploit resources from harsh environments such as deserts, the polar regions (not Antarctica at present but will that have to change?) and ever deeper waters on the continental shelves?
- Are resources sustainable?

> **QUICK CHECK** Divide the consequences of resource exploitation of your chosen resource region into economic and social effects.

2.2 Living with hazardous environments

Key questions for this option:
- What are hazardous environments?
- How can hazards be classified and assessed?

You will be revising:
- Types of hazardous environments
- Living with hazards.

The table at the end of this section shows how you may use different examples for the different generalisations.

Reminder

Do you know Whittow's definition of a hazard?

What is a hazard?

Classifying hazards

There are three types of hazards, However, many hazards overlap.

Magnitude

You will need to know how the magnitude of other hazards, besides these quoted here, are measured.

1 The Fujita-Pearson Scale

Measures intensity of **tornadoes**, the path length and the path width.

2 The Saffir-Simpson Hurricane Scale

The scale based on the hurricane's intensity. It is used to estimate potential property damage and flooding expected along the coast from a hurricane landfall. Wind speed is the determining factor.

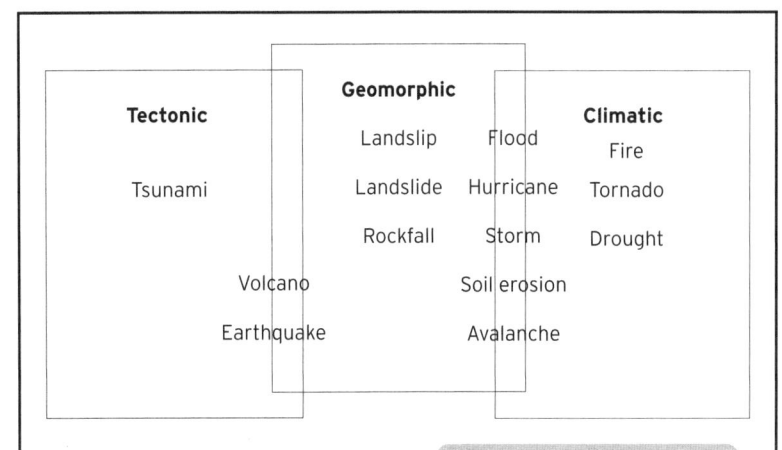

Figure 1 Classifying hazards

- **Tectonic**: Tsunami, Volcano, Earthquake
- **Geomorphic**: Landslip, Landslide, Rockfall, Flood, Hurricane, Storm, Soil erosion, Avalanche
- **Climatic**: Fire, Tornado, Drought

F-Scale	Intensity	Wind speed	Type of damage done
F0	Gale tornado	40–72 mph	Damage to chimneys; breaks branches; pushes over shallow-rooted trees; damages sign boards.
F1	Moderate tornado	73–112 mph	Peels off roofs; mobile homes overturned; moving cars pushed off the roads; garages may be destroyed.
F2	Significant tornado	113–157 mph	Considerable damage. Roofs torn off; mobile homes demolished; large trees snapped or uprooted; light objects airborne.
F3	Severe tornado	158–206 mph	Roof and some walls torn off well constructed houses; trains overturned; most trees in forests uprooted
F4	Devastating tornado	207–260 mph	Well-constructed houses levelled; buildings with weak foundations blown some distance; cars thrown and large objects in flight.
F5	Incredible tornado	261–318 mph	Strong houses lifted off foundations and carried considerable distances; car-sized object fly through the air >100 meters; trees debarked; steel reinforced concrete structures badly damaged.
F6	Inconceivable tornado	319–379 mph	Very unlikely. The small area of damage they might produce would not be recognisable along with the outcome of F4 and F5 wind that surround the F6 winds. Flying cars and refrigerators do serious secondary damage. If this level is ever achieved, evidence might consist of only some type of ground swirl pattern.

Table 1 The Fujita-Pearson Scale

Category One: Winds 119–153 km/hr. Storm surge 1–1.75m above normal. No real damage to building structures. Damage to mobile homes, shrubs, trees and to poorly constructed signs. Some coastal road flooding and minor pier damage. Examples: Allison 1995 and Danny 1997.

Category Two: Winds 154–177 km/hr. Storm surge 1.75–2.6m above normal. Some roofing, door, and window damage. Considerable damage to shrubs with some trees blown down. Considerable damage to mobile homes, poorly constructed signs and piers. Coastal and low-lying escape routes flood 2–4 hours before arrival of the hurricane centre. Small craft in unprotected anchorages break moorings. Examples: Bonnie, North Carolina 1998; Georges, Florida Keys and the Mississippi Gulf Coast 1998.

Category Three: Winds 178–209 km/hr. Storm surge 3–4m above normal. Some structural damage to small houses. Damaged shrubs and large trees blown down. Mobile homes destroyed. Low-lying escape routes cut by rising water 3–5 hours before arrival of the centre of the hurricane. Flooding near coast destroys smaller buildings and damages larger buildings by battering from floating debris. Land lower than 1.75m may be flooded 13km or more inland. Evacuation of low-lying homes close to the shoreline may be required. Examples: Roxanne, Yucatan Peninsula, Mexico 1995; Fran, N Carolina 1996.

Category Four: Winds 210–249 km/hr. Storm surge 4–6m above normal. Extensive end wall and some complete roof structure collapses. Shrubs, trees, and all signs are blown down. Complete destruction of mobile homes. Extensive damage to doors and windows. Low-lying escape routes cut by rising water 3–5 hours before arrival of the eye. Major damage to lower floors of buildings near the shore. Land below 3m may be flooded requiring evacuation of residential areas as far as 1km inland. Examples: Luis over the Islands, 1995; Felix and Opal 1995.

Category Five: Winds >249 km/hr. Storm surge >5–6m above normal. Complete roof removal on many houses and industrial buildings. Some complete building collapses with small buildings blown away. All shrubs, trees, and signs blown down. Complete destruction of mobile homes. Severe window and door damage. Low-lying escape routes cut by rising water 3–5 hours before arrival of the eye of the hurricane. Major damage to lower floors of all structures located less than 5m and within 500m of the shoreline. Massive evacuation of residential areas on low ground within 8–16km of the shoreline may be required. Examples: Mitch over the western Caribbean, 1998; Gilbert, 1988 was a Category Five hurricane at peak intensity and one of the strongest Atlantic tropical cyclones on record.

Frequency

The number of occurrences of a particular hazard per year or other time period is the measure of frequency.

Descriptor	Magnitude	Average annually
Great	8 and higher	1
Major	7–7.9	17
Strong	6–6.9	134
Moderate	5–5.9	1319
Light	4–4.9	13 000 (est.)
Minor	3–3.9	130 000 (est.)
Very Minor	2–2.9	1 300 000 (est.)

Table 2 Number of earthquakes world wide 1990–2003

Web link

For more research, go to the following website and enter the express code 1552S: www.heinemann.co.uk/hotlinks

Several million earthquakes occur each year. Many go undetected because they hit remote areas or have very small magnitudes.

2.2 Living with hazardous environments

QUICK CHECK
1. Do you know the principles of other scales of magnitude?
2. Assess the risk that hazards pose to people in countries at different levels of development.

Magnitude Richter scale	1990	2000	2001	2003 To October
8.0 to 9.9	0	1	1	1
7.0 to 7.9	12	14	15	11
6.0 to 6.9	115	158	126	96
5.0 to 5.9	1 635	1 345	1 243	712
4.0 to 4.9	4 493	8 045	8 084	5 456
3.0 to 3.9	2 457	4 784	6 151	5 169
2.0 to 2.9	2 364	3 758	4 162	4 944
1.0 to 1.9	474	1 028	944	1 389
Total	16 612	22 256	23 534	19 926
Deaths (est.)	51 916	231	21 357	2786

Table 3 Number and size of earthquakes 1990–2003

TYPE	GENERALISATION 1	GENERALISATION 2	GENERALISATION 3	GENERALISATION 4
TECTONIC				
EARTHQUAKES	Plate boundaries. Turkey 2003; Kyoto, Japan; Loma Prieta, USA	Multiple NZ; Japan Turkey; California	Algeria 2003	USA vs India
VOLCANOES	St Helen's, USA; Vesuvius, Italy. Subduction zones, hot spots.	Multiple NZ, Japan and Hawaii; Ruiz, Mexico	Pinatubo, Philippines; Krakatoa, Japan; Etna, Italy	Italy vs Rep of Congo
TSUNAMI	(see web link)	Multiple NZ; Hawaii 1975; Peru 1966		Hawaii centre
CLIMATIC				
HURRICANE	Isabel 2003	Caribbean and USA	Mitch; Isobel; Typhoon Rusa 2002	Caribbean; Miami centre
TORNADO	Tornado alley	Selsey, UK; Nebraska, USA 23/6/2003; South Africa	Forecasting centres	Torro project in UK
STORM/HAIL	Relate to 6474 Gale of 25/1/1990	Multiple NZ; South Carolina 8/10/2003	Met Office	Met Office
DROUGHT	Blocking highs. France/UK 2003. Sahel – ITCZ	Sahel; Zimbabwe	Aid and its management	Link to El Niño (6474)
FIRE	Provence, France; Australia (see web link)	Portugal vs USA	Australia; Indonesia	
FLOOD	Antecedent rainfall, and flash floods. Prague, CZ 2002; Ouse, UK	Multiple NZ; Mozambique	Chichester, UK	Mississippi, USA (6471); Thames Barrier
GEOMORPHIC				
AVALANCHE	Climatic causes. Galtur, Austria	Multiple NZ; Rockies, USA; Norway	Chamonix, France; Norway	USFS Avalanche centre. NZ avalanche info.
LAND SLIP/ Slides, Mudflows	Hong Kong; Wanaka NZ	Multiple NZ; Castleton, Derby, UK; Turkmenistan	Queenstown area, NZ	
ROCKFALLS	Newfoundland, Canada	Multiple NZ; Lulworth, UK		Janod web site; Quebec, Canada
SOIL EROSION	South Downs, UK	Multiple NZ; Dust Bowl, USA vs Sahel, Africa	UK; Marlborough, NZ	

Table 4 Examples of possible case studies for generalisations

2.2a The physical processes causing natural hazards

Key questions for this generalisation:
- What are the physical processes that cause hazards?
- What problems do these cause for people?

You will be revising:
- Hazards and physical processes
- Key issues for people.

Key Concepts

The point of energy release is called the **focus**. The depth of the focus is important because shallow earthquakes, 0–70km deep, are responsible for 75 per cent of energy released by quakes and cause most of the damage.

The point on the surface directly above the focus is the **epicentre**.

P waves are shock waves that travel fastest and vibrate in the direction the wave is travelling and affect solids and liquids. These are a brief warning of the S wave shock to follow.

S waves are slower and they vibrate the rock at right angles to the direction the wave is travelling. They do not pass through liquid. These provide the really violent shaking.

L waves are restricted to near the surface and arrive at **seismographs** last although these will produce a series of violent shocks. It is the shocks that cause the hazard.

The greater the **magnitude** the greater the damage, as the Richter and Mercalli scales demonstrate.

The **intensity** generally diminishes away from the epicentre although some studies have shown that surface geology bends the waves and focuses them on areas much as a lens does with light. This distorts the intensity and accounts for some areas being devastated while a block away there is little damage. Northridge 1994 focused the waves on Santa Monica 21 km away as well as on Northridge itself.

There are **secondary hazards** such as **liquefaction** where deposits with a high water content behave like a fluid (Mexico City). **Landslides** and **rock falls** are other secondary hazards.

Tsunami are waves in the oceans caused by submarine earthquakes e.g. following the Krakatoa eruption. In 1883.

To understand the impacts, responses and the issues of natural hazards you need to comprehend the processes that are involved. You do not have to know all the causes of all hazards but you do need to have a range across at least TWO types. Bob Digby's *Global Futures Options Pack* is a useful starting point.

Hazards as physical processes

Earthquakes

This section only looks at the processes behind two hazards. You will need to know the processes causing any hazard that you choose to study.

Figure 1 is a map of the plates in Turkey where 1200 quakes have taken place in three years. All were shallow being within 20km of the surface.

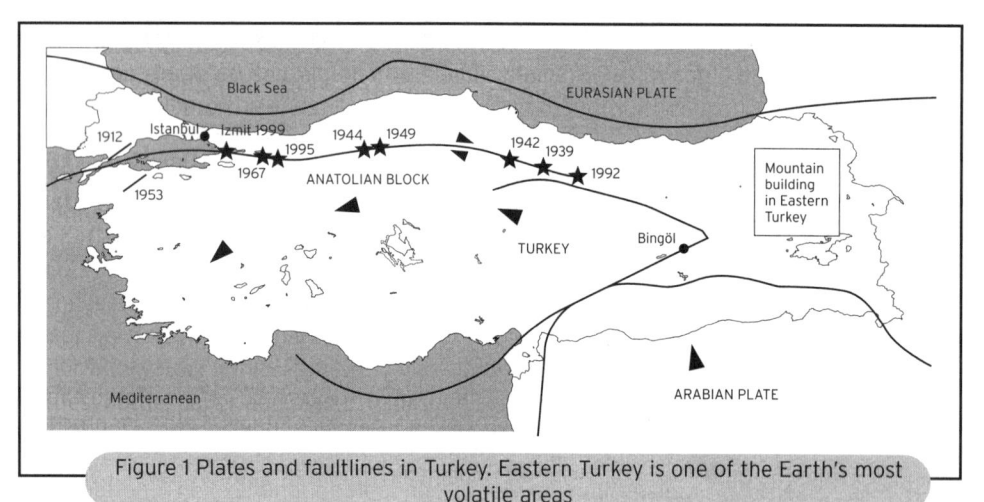

Figure 1 Plates and faultlines in Turkey. Eastern Turkey is one of the Earth's most volatile areas

Tornadoes

They are violent wind-storms speeding across the land at 45km/hour for up to 30km. The spiralling column of air is on average between 150 and 6000m across.

- In the USA they are found mainly in Kansas and Oklahoma – Tornado Alley. They can be found in India, South Africa, much of east Europe and the former Soviet Union, and even occasionally in the UK, e.g. Selsey, 2002.
- In the US complex depressions near the Rockies draw in cool, dry air from the north west which meets warm air drawn north from the Gulf of Mexico.
- These create **super cells** (large, powerful thunderstorms) that spawn tornadoes and require strong **vertical wind shear** (changes in wind speed and direction with height), and **instability.**
  Tornadoes can strike at any time of day, mainly in spring and summer, but are more frequent in the afternoon and evening, after the heat of the day has produced the hot air that powers a thunderstorm.
- A tornadic thunderstorm forms where moist, warm air gets trapped beneath warm, dry air under a capping stable layer of cold, dry air. This air sandwich is called an **inversion.** If the cap is disturbed by a cold front, the warm, moist air can

Reminder

You should be able to relate these processes to one or two earthquakes such as Kobe or Izmit.

2.2a The physical processes causing natural hazards

punch through the stable air above it. Condensing of water vapour releases latent heat and the warm air starts to spiral upward. Aided by different winds at different levels in the atmosphere, the rotating updraft gains velocity. Scientists do not fully understand the first stage of their formation.

- They come in swarms, e.g. tornadoes ran across Tornado Alley (Figure 2) on 4 May 2003 and killed at least 39 people.

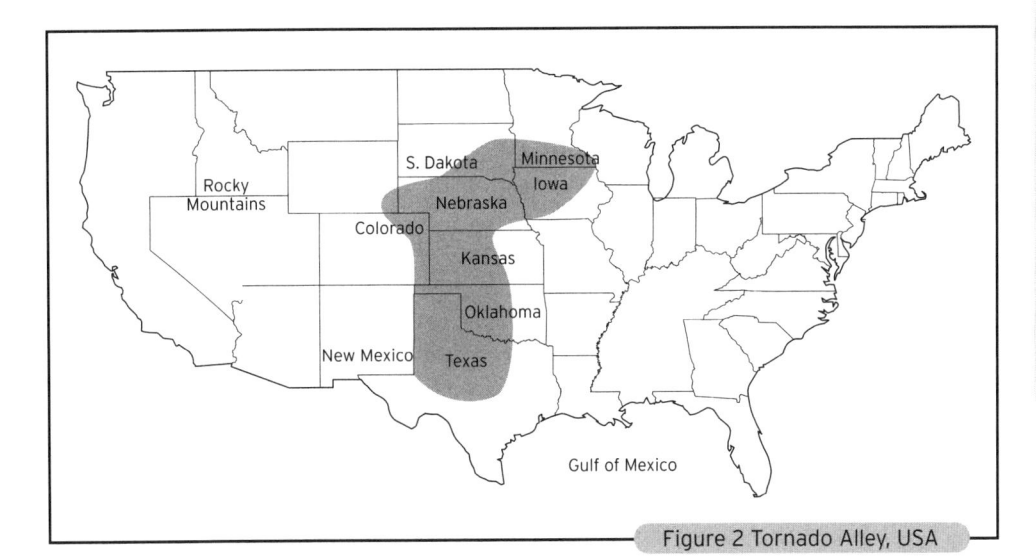

Figure 2 Tornado Alley, USA

> **Reminder**
>
> All your selected hazards can only be understood if you understand the physical processes like those outlined above. Make sure you know the physical processes that cause hurricanes, avalanches, landslides and soil erosion. It is good to have knowledge of processes drawn from the different types of hazards so that you can display breadth of knowledge and understanding

QUICK CHECK

1. Are you able to draw the major plate boundaries on a world map such as the one issued in the examination? Can you mark the divergent, collision, subduction and transform or conservative fault margins on the map? Can you classify the fault margins as constructive and destructive?

2. Can you show how volcanoes link to tectonic processes by distinguishing the types of lava, the types of cones, types of eruption and their location in relation to the plate boundaries?

2.2b The impact of hazards on people, the economy and the environment

Key questions for this generalisation:

- What are the impacts of hazards upon people, the economy and the environment?
- How and why do these vary spatially?

You will be revising:

- The social, economic and environmental impacts of hazardous environments.

The generalisation for 2.4 asks you to research three types of impact and to focus on multiple impacts. The example of multiple hazards selected here is New Zealand although it could easily be Japan or Hawaii. The real impact on an area also depends on the level of development of the affected area.

The impacts of hazards

Volcanoes and volcanic areas

1 Social

- They have a place in the religious rites of peoples, e.g. the Maori people.
- They can destroy communities and cause total social disruption, e.g. Mount Tarawera, New Zealand, 1886, an **explosive** eruption that destroyed whole villages such as Te Wairoa and killed 153 people; **Plinian,** extremely explosive, eruptions such as that from Vesuvius which destroyed Pompeii; the Mt Pelée eruption that killed all but one in St Pierre, Martinique in 1902.
- They often forge closer communities who band together in times of eruption and threat.
- They can cause displacement of people, e.g. Mount Nyiragongo, Democratic Republic of the Congo (Zaire); and Montserrat. In the latter case people are unable to return because their settlements have been buried under lava and volcanic dust, just as happened to Pompeii and Herculaneum in the Bay of Naples.

2 Economic

Volcanoes and volcanic landscapes can:
- Damage tourism and recreation, e.g. the ski centres on Mount Etna.
- Aid tourism, e.g. Te Wairoa, the buried village near Mount Tarawera in New Zealand; Stromboli, Italy to witness its constant outpourings; and to Hawaii.
- Encourage tourism associated with geysers and thermal landscapes e.g. around geothermal fields of Rotorua. Similarly geysers attract tourists to Iceland.
- Provide settings for films, e.g. *The Lord of the Rings* trilogy features several New Zealand volcanoes as a back drop.
- Improve agriculture. The soils around volcanoes are rich once the lava has been eroded. The slopes of Etna and Vesuvius have long had vineyards and farms taking advantage of the soils.
- Provide thermal power, e.g. Wairaki, New Zealand, and Larderello, Italy.
- Provide health spas and treatments, e.g. Volcano, Italy; Rotorua, New Zealand; Blue Lagoon, Iceland.

3 Environmental

Some of the consequences include:
- Deposits of ash and dust, e.g. Tarawera deposited ash and dust 'like grey English snow' across North Island from Rotorua to Hawkes Bay.
- Reshaping of the landscape by lava flows, volcanic mudflows (**lahars**), ash fall (**tephra**) and cinder fields and pumice fields, e.g. the ash fall that overwhelmed Herculaneum.
- Explosive eruptions which can lead to **nuées ardentes**, rapidly moving surface clouds of hot gases (see Plinian eruptions above).

2.2b The impact of hazards on people, the economy and the environment

- Creation of landscape for scientific research, e.g. Surtsey, Iceland.
- **Tsunami,** which can occur from underwater eruptions such as Krakatoa 1883 when the wave encircled the globe and took seven hours to reach Greenwich. So big was the wave that a boat was lifted 60 feet above sea level in Sumatra.

Drought

This case study of impacts is a good one for noting the differences between MEDCs and LEDCs. Droughts in LEDCs are a matter of years whereas in the UK it is a matter of months.

1 MEDCs, e.g. summer and autumn of 2003 in France, Germany and UK.

Social impact

- Need for water conservation. By October 2003 requests to conserve water in some regions regarded as a social duty, included water-saving measures such as 'hippos' in toilets.
- Threat to life when combined with heat, e.g. in France 4000 more deaths than expected in August 2003.
- Pleasure from increased autumn colouring to woodland.
- Low water levels in recreational streams and lakes reduced fishing and pleasure uses.

Economic impact

- Mixed harvests from **agricultural drought** (when the amount of moisture in the soil no longer meets the needs of a particular crop). Poor harvest of some crops, e.g. maize, but good harvest of wine grapes, even in UK.
- Increased cost of irrigating crops.
- Limited ploughing as land too hard to plough for next year's crops.
- Limited seeding as land too hard to seed and seeds not germinating.
- Altered people's purchasing habits over summer period when combined with heat – reduced sales of clothing and increased beer sales.
- Lowered river levels meant that barges were unable to operate fully laden on the Rhine.

Environmental impact

- Dried out pasture.
- Glacial retreat accelerated by high temperatures.
- **Hydrological drought** (occurs when surface and subsurface water supplies are below normal). River flows lowest for many years.

2 LEDCs, e.g. The Sahel

Rainfall in the region has declined by between 20 per cent and 50 per cent, and has led to severe droughts in 1972, 1975, 1984 and 1985.

Social impact

Socio-economic drought is when physical water shortage begins to affect people. It can lead to:

- Starvation and death because wells run dry and crops do not grow
- Increased migration
- Increased Aid through disaster aid at time of crisis and e.g. Water Aid
- Increased research, mainly based in MEDCs, e.g. Drylands Research Centre, London and Dakar and Oxfam's Arid Lands information network
- Pressure to change property rights in the region to benefit the indigenous population.

Economic impact

- Loss of crops and livestock that are the cause of overgrazing
- Seeds for next season's crops do not germinate
- Lack of water for irrigation and loss of food crops and irrigated cash crops such as cotton used to pay debt burden.

Environmental impact

- Lack of rain because ITCZ does not migrate far enough northwards and bring rains
- Loss of bushes that provide firewood
- Dust storms that remove soil and transport it up to 4000km away
- Expansion of desert and semi-desert southwards in West Africa
- Reduced river flow
- Continuing dryness as following years on the savannas are also dry due to enhanced drying out and therefore less evapotranspiration until the losses made up.

> **QUICK CHECK**
> 1 Can you group the social, economic and environmental effects of volcanoes into costs and benefits?
> 2 Work out your own impact study of other hazards.

2.2c How people respond to and manage natural hazards

Key question for this generalisation:

○ How successfully do people respond to and attempt to manage the impact of natural hazards?

You will be revising:

○ Reactive and proactive responses to hazards.

Your research should focus on how people in different parts of the world respond to natural hazards. Responses will depend on where the hazard takes place, its magnitude, and how often it occurs.

Responses to hazards

Avalanches

○ There are 19 avalanche research centres in North America which focus on understanding the physics of avalanches and the most common locations.
○ The USA, Canada, Switzerland, Austria, Germany, France and New Zealand are all able to afford sophisticated prediction systems and warnings (see below).

Aoraki/Mount Cook 8 October 2003

MODERATE danger is present about the divide, and caution is advised in steep start zones and about ridge tops, on ice slopes above 2000m. Care should be taken as day temperatures warm up and the chance of wet slides on most aspects increases. On aspects below 2000m, a rain-affected snow pack has a LOW danger, but travellers should be aware of start zones at higher elevations

Fiordland 8 October 2003
Avalanche Danger High

Well above average snow cover in alpine areas above bushline for this time of year. Harris Saddle on Routeburn Track 3m plus snow depth. Deep Snow cover exists to below bushline. Continuing storms keeping avalanche danger high. Significant avalanche activity observed throughout Fiordland on all aspects to valley floor. Very large destructive (Size 5) avalanches common during storms. Backcountry travel in Fiordland above bushline is not recommended for at least the next 14 days. Travel in valleys affected by avalanche run-out (eg Milford Track) is also not recommended. This is an unusually late winter. We advise you to delay your spring Fiordland tramping plans until the situation improves.

Figure 1 Examples of New Zealand hazard warnings for skiers

Responses:

○ Educational programmes to make people aware, e.g. Colorado avalanche centre
○ Doing nothing in the short term, e.g. 1000 tourists refused to leave the Galtur area when offered the chance to do so (the 1999 disaster killed 37 people). Continuing off-piste skiing is another dangerous reaction but usually made by those who seek danger (11 died near Galtur in the 2000 season skiing off-piste)
○ Systems to reduce the hazard, e.g. firing guns to dislodge dangerous snowfields
○ Building of snow fences, buildings to divert the avalanche stream and planting of forests.
○ Use of avalanche tunnels over roads and rail tracks to keep transport moving
○ Raising awareness, e.g. the Riverdeep website in Alaska.

Researching global futures: Managing natural environments

> *Be avalanche aware:* avoid avalanche-prone areas; if possible get avalanche survival training; be aware of weather conditions, such as rain, heavy snow, or winds, that can quickly change avalanche conditions; carry an avalanche beacon – a radio transmitter that can help rescuers find a victim before he suffocates or dies of exposure. If you are caught in an avalanche, make swimming motions to try to stay on top of the snow and do your best to move to the side of the avalanche. If you get buried, get your hands in front of your face when you come to.

Figure 2: From the Riverdeep website in Alaska

Who responds?

All reactions and responses to hazards can be classified into before an event, during an event and after an event, which may indeed be before the next event.

	Before	During	After
Government	Education of population. Investing in emergency equipment. Research centres. Put earthquakes onto national curriculum.	Declaring state of emergency either at national or local level.	Providing the army to assist clear up. Emergency funds allocated.
Insurers	Raise premiums in areas of high risk. Insist on precautions before insuring.	Do nothing	Pay out for losses. 2002 disasters estimated to have cost $70 billion Reassess risk.
Planners	Design of quake-proof buildings, and utility provision - gas, electricity, water supplies		Reassess controls on buildings
Civil defence	Education of population. Ensure that workers are qualified to cope with the event.	Placed on standby	Assist in clear up
Relief agencies	Prepare stockpiles of relief equipment. Continuing appeal for funds.	Call up volunteers	Sending in disaster relief teams with dogs after quakes. Appeal for funds
Individuals	Depends on perception of risk. Know procedures for protection such as the best hiding spots.	Flee in panic. Prayer. Go to prepared strong point in building.	Changed perception of event and increased awareness of danger

Table 1 Responding to events – examples relating to earthquakes

What is done to protect, alleviate and prevent hazards?

Not much can be done to prevent some hazards although there are schemes that try to prevent further disasters at the same location and warning systems can always be improved.

- Soil erosion can be prevented to avoid mud slides, e.g. Hong Kong which has had some success at preventing further disasters at the same location.
- Fire can be ameliorated by fire breaks, by regulation concerning litter and smoking in areas of risk. This will not stop lightning strikes.
- Protection schemes such as snow avalanche fences and road tunnels or communal hurricane shelters.
- Research into early warning systems for earthquakes (seismic gap theory) and volcanoes.
- Education, e.g. Earthquake training days in Japan.

QUICK CHECK

1. Try to group the responses to the avalanche hazard above into a) do nothing, b) management, c) engineering, and d) forecasting.
2. Take the responses to hazards above and regroup them into actions: before, during and after an event.
3. Rework Table 1 for earthquakes in LEDCs such as Gujerat or Algeria, or for other hazards in MEDCs and LEDCs.

2.2d Future issues in living with natural hazards

Key questions for this generalisation:

- What are the issues in living with natural hazards in the future?
- Can hazard threats ever be predicted or reduced?

You will be revising:

- Hazards prediction
- Collecting data on hazards.

Are hazards more frequent or are we just better at reporting them, and responding to them? As there are more people living in hazard prone areas, what can be done to reduce the risk of further hazards? The key here is prediction.

Why do we think hazards are increasing?

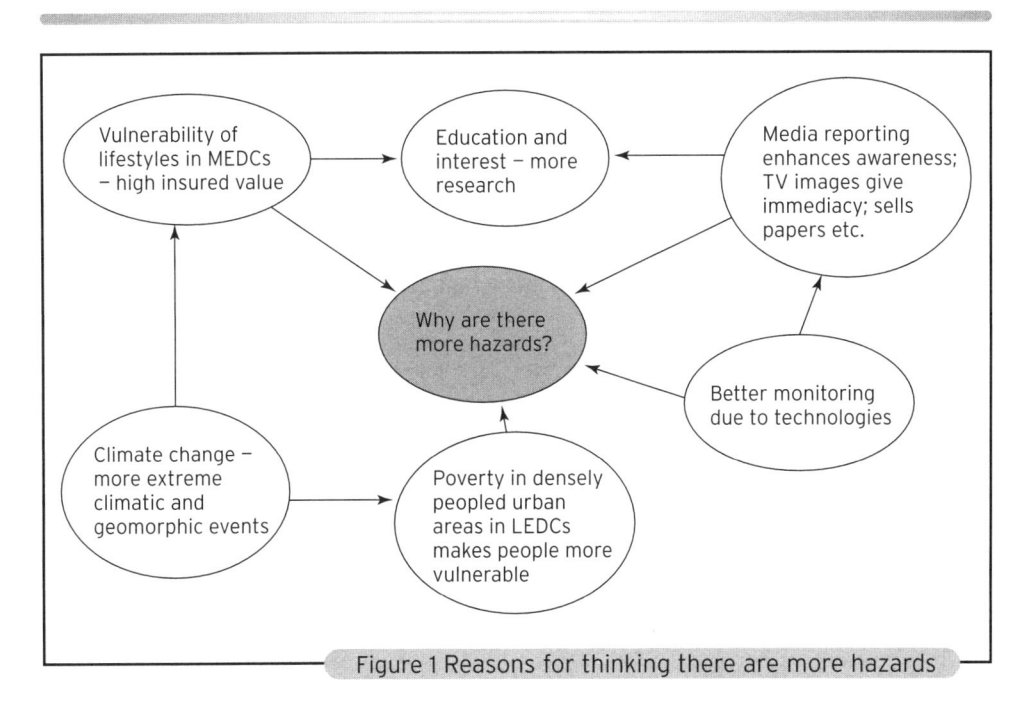

Figure 1 Reasons for thinking there are more hazards

Other questions that must be asked:

1. Are the hazards more frequent or are we just recording every one?
2. Has the physical magnitude of hazards increased causing people to think that there are more?
3. Are more people vulnerable because they live in hazardous regions or has the frequency in these regions increased?
4. Has the media made us more aware because it moves to new news very rapidly? The lifetime of a news story is generally briefer.
5. Are the aid charities more aware of the value of disasters as a means of gaining the vital funds they need? There are more charities seeking our donations.
6. Are more people visiting hazard-prone areas such as ski slopes?

Hazard prediction

Why predict?

- To reduce the effects in terms of loss of life and damage to property.
- To enable people and the economy to recover more rapidly after the event. Very much a MEDC/NIC view because of the high value of investment in factory plants, transport and communications and skills.
- The same view can be expressed for LEDCs although the emphasis is often more rural and linked to survival.
- It is an academic matter that keeps a whole research agenda vibrant.

Examples of prediction

1 Large scale

- Figure 2 shows the global network of hurricane and typhoon warning centres. Some are military because severe weather predictions affect naval vessels. All predictions are shared but the ability of some countries to issue and disseminate warnings is hampered by the available technology. These centres use satellite and radar technology to predict the paths of hurricanes and issue warnings. In the case of Isobel 2003, they got it right, but that is not always the case!

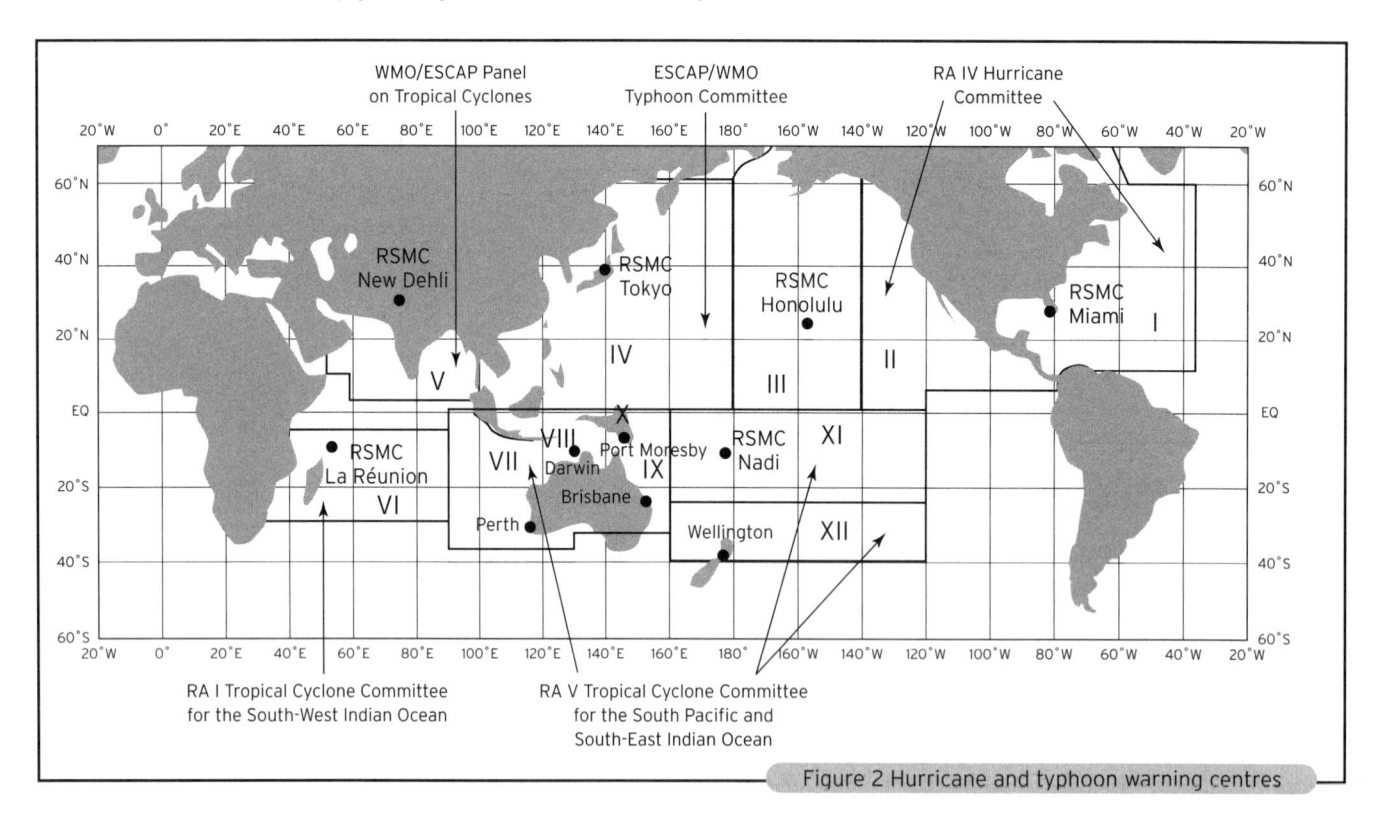

Figure 2 Hurricane and typhoon warning centres

- Pacific Tsunami Warning Centre Hawaii is a large-scale prediction centre. This is supplemented by a separate centre in Palmer for Alaska and West USA. There is an International Centre established by UNESCO attached to the Hawaii Centre. Seismic-monitoring stations and sea-level gauges provide warnings.

2 Smaller, national scale

- Soufriere volcanic eruption on Montserrat in 1995 was predicted by studying earthquakes over the preceding years. The sequence of events followed the predictions made by geologists and enabled safe evacuation of people. Mount Pinatubo and Mount St Helens also involved relatively accurate predictions although some chose to ignore them.

2.2d Future issues in living with natural hazards

- Japanese volcanic predictions began in 1974. Bore-hole seismometers, tiltmeters and strainmeters could detect clear though minor movements of intruded magma even in dormant volcanoes when they are waking up.
- Bush fires in Australia are predictable although the points where they break out are less predictable. Dumped glass bottles and plastics can focus the sun's rays and start fires. The conditions which exacerbated the 2002 fires in New South Wales were temperatures in excess of 35°C, humidity of less than 10 per cent and a wind speed of 40km/hour. These conditions become prevalent when there is a very stationary high pressure system over Eastern Australia in the summer. Vandalism also played a part and its location is not predictable.

> **QUICK CHECK** How effective are we at predicting hazards? List other examples of prediction for, for example, avalanches.

Web link

For more information on drought impact on Africa, go to the following web site and enter the express code 1552S:
www.heinemann.co.uk/hotlinks

2.3 Pollution and natural environments

Key questions for this option:
- What is pollution?
- How can pollution types be classified?
- What is its extent?

You will be revising:
- Types of pollution
- Monitoring pollution.

This option for Global Futures overlaps very strongly with 1.3, 1.4, and 1.5 and the options in 1.6–1.8 of Global Challenge (pp 16–35). The topic asks you to question how we use our environment and what the effects of our activities are. The scale of pollution ranges from local incidents, that you may be able to research, right through to large-scale pollution incidents.

What is pollution?

It is the alteration of the natural environment as both a direct and indirect result of human activities that affects the ability of natural systems to function effectively.

How do we study pollution?

1 **Scale**
- At a point, e.g. light pollution from stadia such as Old Trafford, Manchester.
- In a small area, e.g. lead pollution near a road.
- In a region, e.g. sewage discharge on a small stream.
- In a country, e.g. atmospheric pollution in Poland
- At a continental scale, e.g. pollution on the Rhine, acid rain.
- Global scale is normally the effects of, e.g. the enhanced greenhouse effect.

2 **Timescale**
- Over a short period, e.g. noise and light pollution at Fratton Park, Portsmouth, or Old Trafford, Manchester. Even cricket now has floodlit matches.
- Over a period of 1 week to a month, e.g. smog over Athens.
- Over a long period, e.g. oil spills, atomic radiation fallout.
- Continuous, e.g. acid rain, seepage from waste dumps.

3 By its three components:
   (a) Its source or sources
   (b) Its pathway – where does it spread to?
   (c) Its sinks – where does it end up?

4 By linking it to human activities: industry, transport, household activities, events. Do some of these activities cause more pollution than others?

5 By relating it to levels of development. Are pollution events different in LEDCs to MEDCs? Are the events different in the former Socialist states?

Monitoring pollution

Who monitors and measures?

1 The individual by perception of smell, views, water quality, landscape
2 The company trying to maintain a good environmental protection record, e.g. The Body Shop
3 The local authority charged by government with enforcing legislation
4 Area-specific administrative bodies, e.g. Chichester Harbour Conservancy, or a National Park Authority

2.3 Pollution and natural environments

5 National government that passes the laws
6 The International Community through UNEP and other treaties
7 Pressure groups at all levels from local to global.

The table below gives examples of case studies for each generalisation.

Type	Generalisation 1	Generalisation 2	Generalisation 3	Generalisation 4
Atmospheric				Kyoto; den Haag; Johannesburg
Traffic	Local – lead pollution, SO_2, nitrogen oxides	Disease; Quality of Life (QoL); corrosion	Motorists, noise pollution	Congestion charging
Smoke	Local – warehouse fire and toxic fumes. Regional – the haze in Indonesia etc.	Disease; health issues	Households and industry	Clean Air Acts
Smog	Regional – Los Angeles, Athens	Disease; QoL in LA; Corrosion	Motorists	Clean Air Acts, Electric and hybrid cars
Ozone layer	Halon and CFCs			
Chemical clouds including radioactive.	National to International – Chernobyl	Demographic and health; Clear up costs.	Industry; Greenpeace, Friends of the Earth	International treaties
Biosphere				Earth summit
Acid rain	Regional to International	Agricultural, forestry and building erosion costs	Motorists; power generation and industry	International agreements and target setting.
Hydrosphere				Earth Summit, Johannesburg
Rivers	Local to International – Sandoz chemical spill.	Costs of water supply; Loss of fishing; Costs of treatment	Role of agriculture, sewage/water companies	International waterways, e.g. Rhine
Oceans	National to international – Oil spills	Milford Haven, loss of fishing	Shipping companies; Greenpeace	International agreements
Land				Earth summit
Derelict land	Local to regional – Mining	Costs – asbestosis; QoL	Role of government; property developers	Government initiatives
Waste dumps	Local case study	QoL; disease	Role of consumers; local government	Legislation; Agenda 2; recycling

Table 1 The case study grid for pollution

QUICK CHECK — What are the main pollution concerns in your home area?

2.3a Measuring variations in pollution

Key question for this generalisation:
- Why does pollution vary spatially and over time?

You will be revising:
- Scales of pollution
- Pollution and economic development.

The key theme is the linkage between economic development and pollution. All activities can be polluting. At different levels of development, different types of pollution are more common.

Scales of pollution

Technological advances have led to more research into scales of pollution. Pollutants are accumulating in marine sediments, on glaciers and ice caps and in the fat of arctic animals. Some examples of pollution include:

- Marine pollution: 75 per cent originates on the land. Sewage is the main offender. Impact is relatively confined in streams but in oceans the stream flows merge to affect larger areas. Nitrates entering the North Sea have risen by 400 per cent between 1970 and 2000. Very often a single source may seem relatively minor but when there are multiple sources, the effects can be very serious.
- Acid rain is the accumulated product of many power stations and not just one.
- Domestic waste: One household's waste amounts to a few tonnes per year but a population of 10 000 results in a large demand on disposal sites, especially if none of them uses recycling.

Scale can be measured by the cost of cleaning up the pollution, the area affected, the number of people affected ranging from those sheltering from toxic fumes from a fire to thousands affected by a nuclear radiation fall-out.

Figure 1 shows how Halon, a gas associated with CFCs in damaging the ozone layer, has vanished from MEDCs but China has increased its use.

Pollution and economic development

Improved technology is a prime cause for the 75 per cent growth in pollution and that comes with development. Some say that cleaner technologies suggest that technology is not the main culprit. However, pollution is cumulative and past excesses are still apparent. Causes of pollution include:

- Population growth, which accounts for approximately 25 per cent of growth in pollution in the USA
- Wealth and the use of the internal combustion engine. Figure 2 shows CO_2 production in 1998.
- Chemical applications to fields, which have increased especially in countries that experienced the Green Revolution. Over half of the fertilisers used have evaporated or been washed into rivers. Therefore, in both Asia and Europe where population pressures are highest, nitrogen pollution and eutrophication is highest. Off the coasts **algal blooms** result that shade out shellfish as in 1996 in West Scotland; in 1998 in California and in the Adriatic.

2.3a Measuring variations in pollution

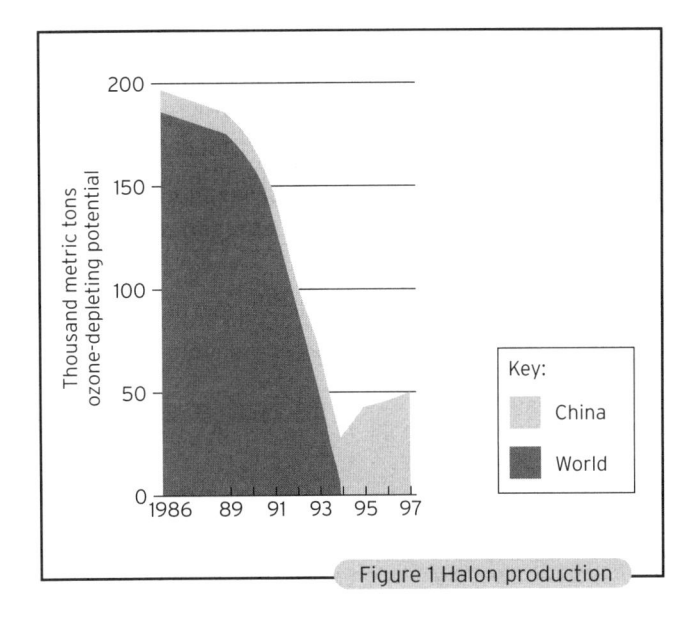

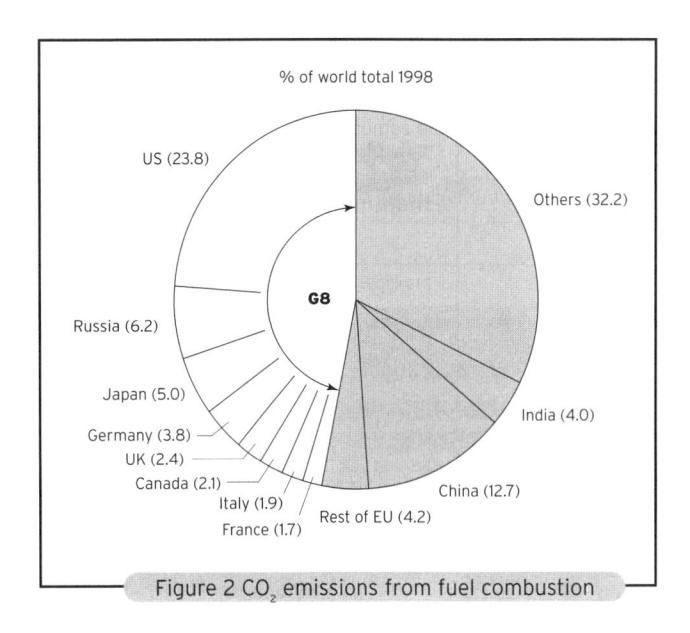

Figure 1 Halon production

Figure 2 CO_2 emissions from fuel combustion (% of world total 1998): US (23.8), Russia (6.2), Japan (5.0), Germany (3.8), UK (2.4), Canada (2.1), Italy (1.9), France (1.7), Rest of EU (4.2), China (12.7), India (4.0), Others (32.2). G8.

- Pesticide usage, which has risen by over 300 per cent since 1950. It is associated with advanced agricultural practice and with intensive cash crop farming in LEDCs. Insects and diseases become resistant – some areas moving to organic farming as a result.
- Incidents at offshore oil and gas rigs. Drilling for oil and gas is generally safe but can have major effects if an incident occurs. It is the demands of the developed world for oil that lie at the heart of the drive to exploit ever more dangerous seas. There are 500 production platforms in the North Sea.

Why is there more pollution?

- We are better at reporting it. Technologies enable monitoring.
- Lack of effective controls in the past. CO_2 levels have risen by nearly 80 per cent since pre-industrial times. Methane concentrations in the atmosphere have risen 146 per cent in the same period.
- In MEDCs emissions that were not visible were not subject to control.
- Some products are new and their ability to pollute not noted until too late, e.g. chlorines in the atmosphere.
- Use of cars is rising. Air travel is rising. People in MEDCs expect to move about, commute, have weekend breaks, have two or three cars, go on holidays by air.
- Our economies are increasingly dependent on energy for powering our homes, running computers, producing goods, e.g. robots. Witness what happened to New York, London, Italy, Denmark and Sweden when power failed in 2003. All demand energy which involves the conversion of oil, gas and coal to electricity, nuclear power (70 per cent of French electricity is from nuclear sources). Emissions from power stations are declining but this is partly offset by increasing demand especially in NICs.

City	Sulphur dioxide (microg/m³)
Guiyang	424
Tehran	209
Turin	159
Brussels	129
Istanbul	120
Moscow	109
Mexico City	74
Berlin	50
Toronto	36
London	25
Auckland	3

Table 1 Air pollution in a selection of cities

Researching global futures: Managing natural environments

○ Rapidly industrialising countries such as China depend on old technologies such as coal power. Chongqing Green Volunteers League have been protesting about toxic waste dumping from a plant at Xiao'anxi into the Yangste. The city and national government are not acting because the Chinese priority is economic growth. Illegal dumping is a major concern as the waters of the three Gorges dam build up holding back the pollutants.

QUICK CHECK

1. Can you offer any explanations for the data in Table 1?
2. Is there more pollution or are we better at monitoring it and, by legislating against it, creating the bodies who monitor and record pollution?

2.3b Environmental, social and economic impacts of pollution

Key question for this generalisation:

- What are the environmental, social and economic impacts of pollution?

You will be revising:

- The environmental, social and economic impacts of pollution.

You should understand the types of pollution and their relationship with economic development. You should have case studies of all types of impact at different scales from the local to the global.

Environmental impacts

Environmental impacts will often affect ecosystems and the built environment. Much depends on the type. They include:

- Smoke has blackened buildings and corroded stone from the nineteenth century.
- Traffic fumes have also had the effect of increasing chemical weathering of limestone building stone.
- Smog reduces visibility from viewpoints such as from Likavittos across Athens to the Acropolis or the Hollywood Hills across Los Angeles. It also damages plant life.
- Chemical clouds affect air quality and lead to breathing difficulties and possibly kill vegetation.
- Acid rain results in vegetation stress, defoliation and kills freshwater fish.
- River pollution often leads to the death of fish and water plants and the subsequent food supply problems further up the food chain. It also means that the water cannot be used by people without further purification.
- Ocean pollution results in the growth of algal blooms; the death of sea plants, e.g. sea grasses; loss of shellfish and fish and less food for their predators. Oil spills can have a devastating effect on the environment, such as Exxon Valdez when 40 000 tonnes spread along a fragile coastal ecosystem. The effects on ocean deep are not really known.
- Derelict land encourages fly-tipping and is often unsightly.
- Waste dumps encourage vermin, create smell pollution in warm weather, and can affect underground water. Later there may be a build up of methane which has to be extracted safely, e.g. the landfill site, Port Solent, Portsmouth Harbour has small chimney fans scattered over the site.

Reminder

Always try to have examples from more than one type of pollution in and from a variety of scales.

Social impacts

The type of pollution will determine the scale of impact.

Pollution	Impacts
Traffic	○ Health – affected by noise; leaded petrol damages children's brain development; particulates from diesel engines damage lungs ○ Poorer quality of life, e.g. proximity to motorways, traffic jams
Smoke	○ Health – particularly those suffering from asthma ○ Antisocial aspect of bonfires ○ Toxic fumes from warehouse fires where chemicals are housed
Smog	○ Health – can cause eye irritation, asthma; powerful irritant, especially for elderly, very young and sufferers from cardiovascular or respiratory illness ○ Can cause death, e.g. London in 1950s
Ozone layer	○ Health – holes in ozone layer can cause malignant melanoma
Chemical and radioactive	○ Health – radioactive areas closed; in some cases fertility affected; radiation disease and cancers
Acid rain	○ Recreational and tourist landscapes spoilt ○ Fishing loss ○ Future income loss from forests for workers
Rivers	○ Social use of rivers interrupted ○ Introduction of nitrates into water supply ○ Disease where river used for washing and water supply
Oceans	○ Tourist beaches and coastlines made inaccessible ○ Fishing industry – loss in catches especially crustaceans
Derelict land	○ Anti-social behaviour encouraged
Waste dumps	○ Property prices near dumps or potential dumps decline ○ Methane accumulates and can be dangerous

Table 1 Social impacts of pollution

Economic impacts

Costs of pollution include:

○ Cost of clean up, e.g. the cost to the company – Exxon in the case of Exxon Valdez oilspill; cost to Sandoz in the case of the 30 tonne chemical spill into the Rhine at Basel
○ Cost in human lives and human health, e.g. the Union Carbide factory at Bhopal is still fighting law suits to prevent huge costs of the 1984 explosion; the AZF explosion in Toulouse in 2001 killed 30 and maimed others
○ Health costs, e.g. birth defects have followed the spread of the chemical cloud at Bhopal. In a sample of 865 women who lived within 1km of the plant and who were pregnant at the time of the gas leak, 43 per cent of the pregnancies did not result in live births. Of the 486 live births, 14 per cent of babies died in the first 30 days compared to a death rate of 2.6 to 3 per cent for previous deliveries in the two years preceding the accident in the same group of women. 'Children are being born with deformities like cleft palate, three eyes, all fingers joined, one extra finger, one testicle, different skull shapes, and Down's syndrome'. (Source: The Lancet, 14 September 2002)

> **Reminder**
>
> A pollution incident is a single event from a single source.

2.3b Environmental, social and economic impacts of pollution

Cost of measures to reduce pollutants, e.g. scrubbers to power station chimneys, catalytic converters in cars, improved water purification plants to cope with detergents in sewage
- Cost of sustained pollution especially where the source is increasing. Coping with car pollutants is progressing but more cars on the road reduce the overall effects of the measures. Also sustained effects from radioactive accidents such as Chernobyl
- Loss of income from affected resources such as forests, fishing.

Most of the impacts cannot be seen in isolation. There will always be a combination of social, economic and environmental.

Impacts of the Sea Empress oil spill, 15 February 1996

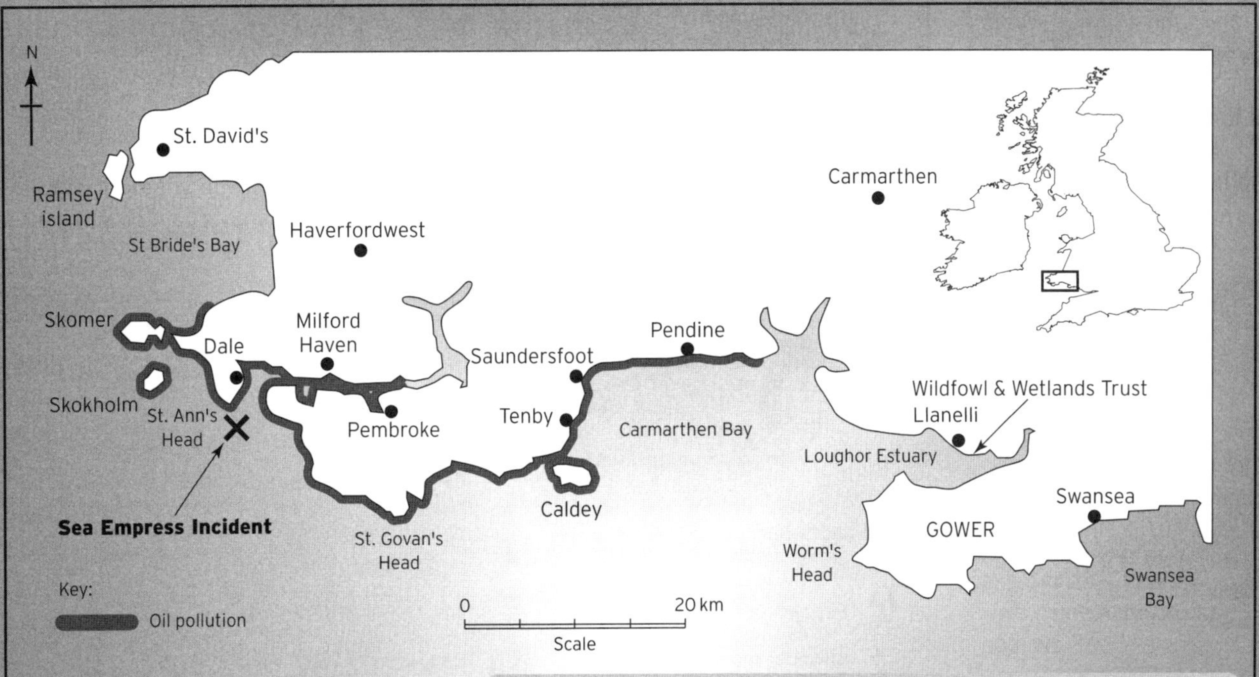

Figure 1 Extent of heavy oil pollution from the Sea Empress incident

The scale of pollution (70 000 tonnes of crude oil) approached that of the *Torrey Canyon* which spilt 117 000 tonnes of oil around Cornwall in 1967. In 1978, the *Amoco Cadiz* was wrecked off Brittany – the world's worst oil tanker spill of 223 000 tonnes of crude oil.

During the early weeks, oil was observed across a wide area of the Bristol Channel including Lundy Island and the south east of Ireland. The clean up included:

- Aircraft used to spray chemical dispersants onto slicks at sea
- Specialized vessels used to recover oil from the sea surface
- Booms to protect some ecologically sensitive areas
- A large workforce during the onshore clean-up operation; hand scrapers, mechanical diggers used on sandy shores
- Scrubbing rocks, applying chemicals followed by high pressure water jets, and using suction on heavily oiled shores.

Many inaccessible shores remained uncleaned. Natural processes played the major role in the clearance of these shores. Natural cleansing was rapid for many exposed rocky shores, but was much slower for sheltered, vegetated muddy shores. However, natural dispersal and intensive clean-up activity restored the aesthetic appeal of badly affected shores by April 1996 and, by the summer, bathing and watersports had recommenced.

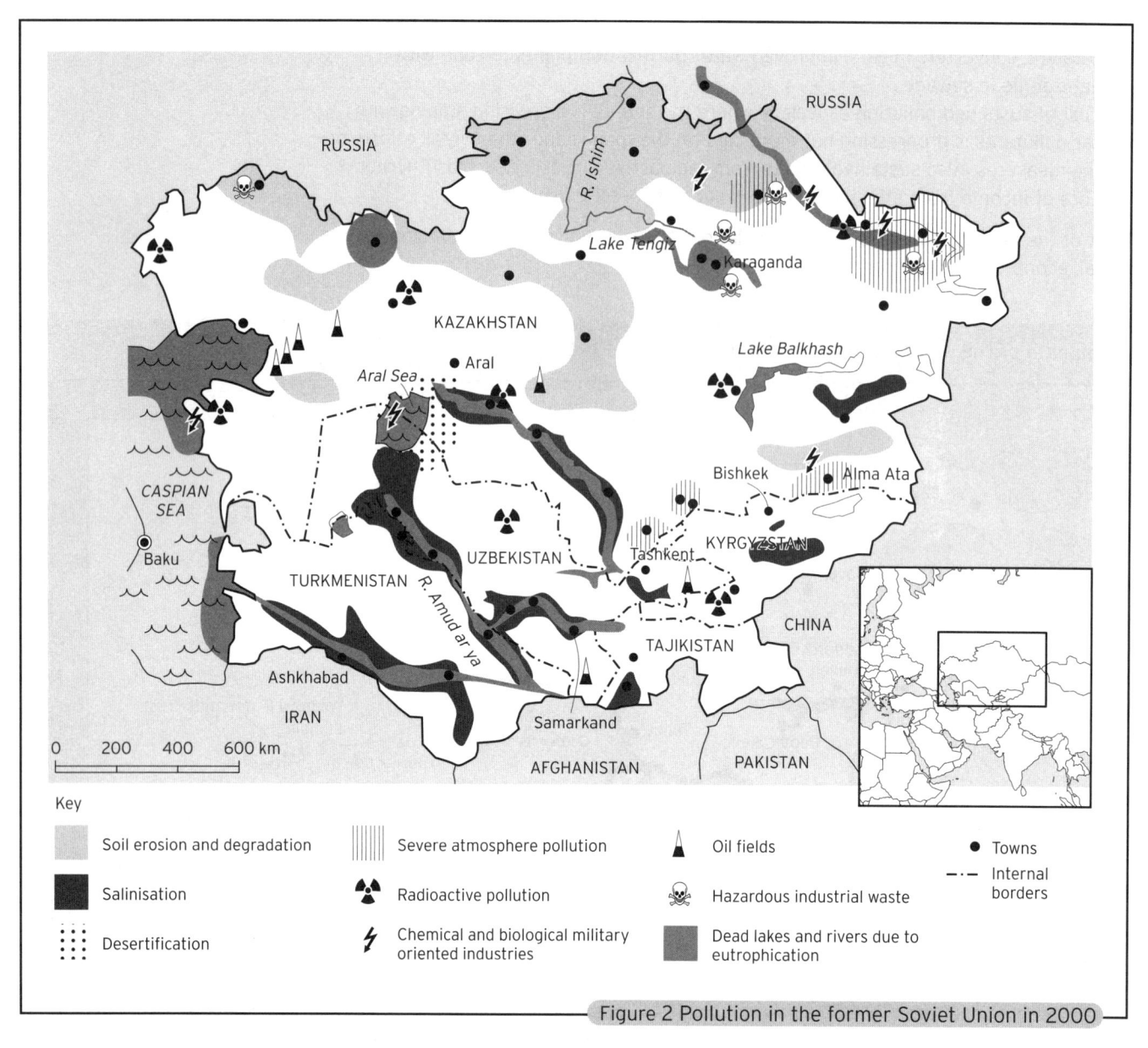

Researching global futures: Managing natural environments

Key

- Soil erosion and degradation
- Salinisation
- Desertification
- Severe atmosphere pollution
- Radioactive pollution
- Chemical and biological military oriented industries
- Oil fields
- Hazardous industrial waste
- Dead lakes and rivers due to eutrophication
- Towns
- Internal borders

Figure 2 Pollution in the former Soviet Union in 2000

The map in Figure 2 shows multiple pollution that exists out of sight in the former Soviet Union and is rarely commented upon by the Western media.

QUICK CHECK

1. Are the environmental impacts the same in LEDCs as in MEDCs? Give reasons for your opinions.
2. Which of the social effects in Table 1 are long term and which short term?
3. Explain the economic origin of the forms of pollution in Figure 2. What can be done to solve the problems created and what might be the effects of solving the problems?
4. Why is pollution in the former Soviet Union not given more prominence.

2.3c Managing pollution incidents

Key questions for this generalisation:
- How can pollution incidents be managed?
- By whom should they be managed?

You will be revising:
- Human and natural causes of pollution
- How to manage them.

The Encyclopaedia of Environmental Pollution and Cleanup *is a useful technical resource for this section. The key is to research whether the polluters themselves are the ones taking corrective initiatives or whether it is governments or pressure groups that successfully bring about action.*

Human and natural causes of pollution

Who are the polluters?

For any pollution type it is essential to consider the range of potential sources of an incident.

Possible human causes of river pollution:

- Builders allowing heavy run-off from sites to get into rivers
- Malfunctioning waste water, sewage plants
- Industrial spillage
- Industrial discharges, both legal and illegal
- Waste water from mines
- Antisocial behaviour such as sump oil being deposited in drains
- Farmers over applying nitrogen fertilisers
- Slurry from animal pens. Other wastes such as chicken manure
- Oil leaks from pleasure boats
- Waste being thrown overboard
- Illegal fly-tipping on banks
- Untreated sewage. (City populations who still do not have access to improved sanitation: Asia 74 per cent, Latin America 13 per cent, Africa 11 per cent, Europe 2 per cent and North America 0 per cent.)
- Cultural norms such as cremations on the banks of the Ganges.

Possible natural causes of river pollution:

- Flash flooding might be a natural cause of pollution from urban areas because systems cannot cope with the increased flow.
- Excessive erosion of soils may be seen as natural although it is possible to blame agricultural practices such as ploughing up and down a slope.

The table below shows some sources of organic water pollution in a range of countries.

Country	Kg/day	Food and beverages	Paper and pulp	Metals	Textiles	Chemicals
UK	611 743	33%	28%	8%	7%	12%
China	8 491 856	30%	12%	17%	15%	13%
Germany	811 316	31%	17%	13%	5%	16%
Indonesia	347 083	33%	19%	5%	23%	10%
Sierra Leone	4 170	82%	10%	0%	2%	3%
Tunisia	186 269	43%	7%	11%	26%	8%

Table 1 Sources of water pollution

Who monitors pollution?

- Pressure groups and charities such as International Fund for Animal Welfare (IFAW), Greenpeace, Friends of the Earth.
- Governmental agencies and departments such as the Environment Agency, the Department for Environment, Food and Rural Affairs (DEFRA), National Rivers Authority (NRA).

The management of some types of pollution

The rash of oil spills, e.g. Nadhodka, Japan, 1997; Pallas, Germany, 1998; Erika, France, 2000; Treasure, South Africa, 2000; Bunga Teratai Satu ship grounding, Australia, 2000; Levoli Sun chemical ship sinking, France, 2000, all emphasise the need for regulatory legislation to protect the environment and hold the responsible party accountable.

Some examples of legislation against oil pollution include:

- At the international level the requirement for oil tankers to have double hulls, introduced in the US Oil Pollution Act 1990 after the Exxon Valdez spill has been copied in Europe. It has not yet stopped spills. There is no international standard although countries have introduced anti-spill legislation.
- California taxes all oil that passes through its waters. This funds conservation, building of facilities and the strengthening of an oil–wildlife network in California.

Some case studies of managing pollution include:

1 Fly-tipping

Managing the consequences of fly-tipping has become an increasing problem in recent years. To help determine the nature and extent of the problem on agricultural land Tamar Consulting was commissioned to design a survey of 400 farmers in England and Wales. The survey was supported by the National Fly-tipping Stakeholders' Forum, and funded by the Environment Agency and DEFRA. As a result of this cooperation between an NGO, a research consultancy and the government, measures were introduced to reduce pollution.

2 Smog

Smog Forecasts in the Maritime Provinces of Canada is a successful government-backed management scheme to alleviate the effects of pollution. Environment Canada's twice-a-day smog forecast is now available as part of the Government of Canada's commitment to provide clean air information. More than 5000 Canadians die prematurely from air pollution-related diseases each year. The Smog Forecast Program in New Brunswick, Nova Scotia and Prince Edward Island, began in 1997. Smog Forecast partners include provincial environment and health departments and the provincial lung associations. It allows people to take predicted smog levels into account when planning their daily activities. The latest Environment Canada Smog Forecast is available on the Internet (see web link). In partnership with government and key stakeholders, Environment Canada introduced a comprehensive Clean Air Agenda. This will reduce harmful air emissions from trans-boundary pollution (i.e. mainly from the USA), the transportation and industrial sectors.

Web link

Go to the following website and enter the express code 1552S:
www.heinemann.co.uk/hotlinks

2.3c Managing pollution incidents

3 Pollution of homes caused by flash floods

Water companies have the statutory authority to prevent and manage pollution but sometimes events work against them. In 2000 flash flooding overwhelmed the drain system in Portsmouth and the local pumping station failed. The consequence was that polluted waters burst up out of the drains and filled former shallow river valleys in the city, flooding properties and polluting the homes and streets. The area was cleared up over a period of three months but the attribution of blame has not been resolved despite the payments to householders by the insurance companies. This incident has been successfully managed in that all have their homes back unpolluted. However, insurance companies and the water company still dispute who is to blame for the event.

4 An unsuccessful case: river pollution

A spill of cyanide at Baia Mare spread through western Romania, Hungary and Serbia. Waste water containing cyanide, from the tailings left from activities of the company Remin, flowed into the Lapus River, and reached the river Danube. Workers at Baia Mare reinforced the dam holding back waste water to prevent more cyanide from spilling into the Lapus River. Baia Mare is owned by companies in Perth, Western Australia and the Romanian government and foreign investors. The company denied responsibility for the spill, and claimed that the extent of the poisoning had been exaggerated. According to the Swiss environment agency it is not possible to assign responsibility for downstream pollution of the Danube. The Australian firm operating Baia Mare will only be liable for damage in Romania. Downstream countries will have no rights to compensation. The International Convention on Trans-boundary Effects of Industrial Accidents and the Convention on the Protection and use of Trans-frontier Watercourses and International Lakes 1992 by the UN Economic Commission for Europe (UN/ECE) had not been ratified by Romania thus making the cost of clean up fall on the victims and not the polluter.

QUICK CHECK

1. Make a list of other potential polluters for atmospheric and land pollution.
2. What makes a management policy for pollution incidents successful?

2.3d Alternative strategies for managing pollution in the future

Key questions for this generalisation:
- **What are the alternative strategies for managing pollution in the future?**
- **What are the social, economic and environmental consequences of each of these strategies?**

You will be revising:
- Pollution strategies
- The future consequences of these.

The theme is ensuring that the responses to pollution incidents are focused on returning the environment to a sustainable equilibrium.

International strategies

The two conventions noted in box 4 on page 105 are examples that failed because countries had not signed up. The problem with all international agreements is that it takes time to get everyone on board.

Examples of international agreements include:

- Central Commission for the Navigation of the Rhine dates back to 1815 although its role in pollution is more recent. The transport of dangerous goods is regulated.
- World Water Forum (The Hague, 2000 and Tokyo, 2003) set up programmes for water and sanitation that often impinge on pollution issues.
- EU legislation and directives have been developed to tackle emissions from transport. For instance, a sulphur protocol was agreed in 1985 and updated in 1994.
- New regulations for twin hulled oil tankers are in force but many of the flags of convenience still permit the older tankers to register with them and continue trading.

The Kyoto Protocol 1997

The Kyoto Protocol is only a first step to achieving the goal to stabilise greenhouse gas concentrations in the atmosphere 'at a level that would prevent dangerous interference with the climate system'. Even if the Protocol is ratified and nations abide by its terms, its effect will only slow – not halt – the build-up of greenhouse gases. Unlike the **Montreal Protocol** (on substances that deplete the ozone layer), which will eventually 'solve' the problem of ozone depletion, Kyoto will not 'solve' the problem of climate change. However, it might begin the long process of arresting the heavy dependence on fossil fuels and other sources of greenhouse gases.

It took until 2002 to get the EU and Japan to ratify the protocol. The USA and Russia have still to sign. By the Johannesburg Earth Summit 'The Bush administration and associated oil and heavy industries took most of the flack for this state of affairs, although all the world's politicians should bear some of the blame.' *The Guardian*, 6 May 2003.

The treaty is unclear about the extent to which developing nations will participate in the effort to limit global emissions. Kyoto made it clear that MEDCs are most responsible for the build up of greenhouse gases and should take the lead in combating climate change. However, LEDCs have a role to play protecting global climate.

Another issue is the practicability of achieving the specified level of emissions reductions (5 per cent off 1990 levels). This task presents a challenge to many countries, because greenhouse gas emissions have grown significantly since 1990 and are projected to continue growing. The US Department of Energy estimates show that by 2010 US carbon emissions will increase 34 per cent from the 1990 level if there are no changes in energy policies and consumer behaviour. The USA, the leading contributor to global greenhouse gas emissions, will therefore need to reduce emissions by more than 33 per cent to stand still.

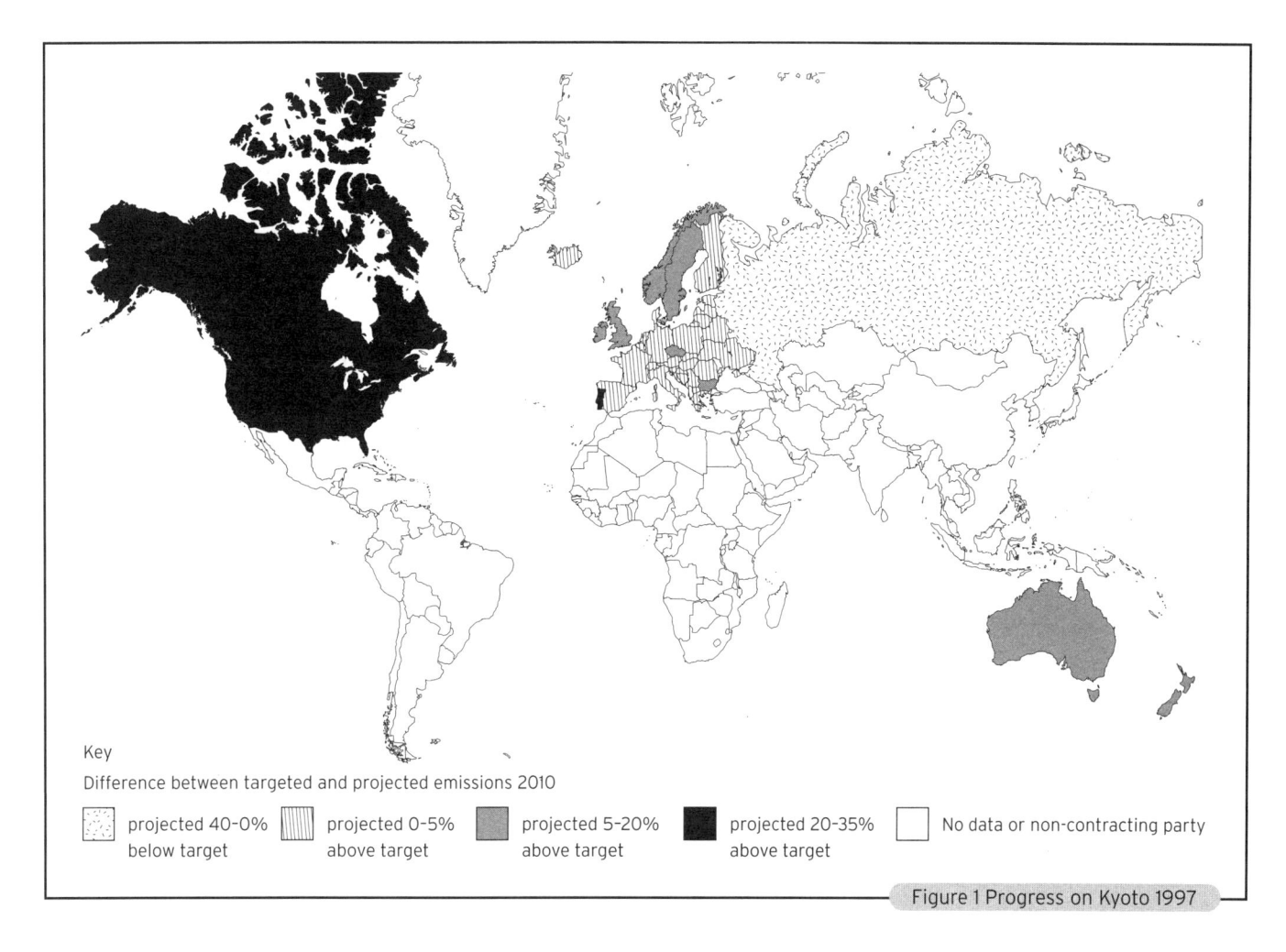

2.3d Alternative strategies for managing pollution in the future

Key
Difference between targeted and projected emissions 2010

- projected 40-0% below target
- projected 0-5% above target
- projected 5-20% above target
- projected 20-35% above target
- No data or non-contracting party

Figure 1 Progress on Kyoto 1997

National strategies

Examples include:

- Altering lifestyles:
  In Germany the government introduced a deposit on drinks bottles and cans in response to the EU recycling directive. The consequence has been that cans have disappeared from sale. Bottles are returnable and more effort has switched to using glass rather than plastic because they are more easily reused and returned. In Japan the cost of recycling a car is built into the new car price, thus making the customer pay for recycling when a product has ended its working life. The obligation is then on the manufacturer to dispose of it.
- Green Consumerism initiatives being pushed by the UN which is trying to get companies to work with the public.
- Legislation that makes the polluter pay.

The two graphs (Figures 2 and 3) show that paper and glass recycling are still in their infancy. Consumption is outstripping efforts to recycle. The graph for glass recycling illustrates the considerable national variations in the early success of the initiative.

Local initiatives

'Think global but act local' is the aim of many schemes and individual crusades and include:

- Improved recycling and not just the green bin. Most people do not put all that could be recycled into the bins, e.g. food containers. Encouraging composting is another local scheme supported by local authorities subsidising composting bins. Raising awareness by providing separate rubbish bins in airports such as Frankfurt where the rubbish is sorted by the public.

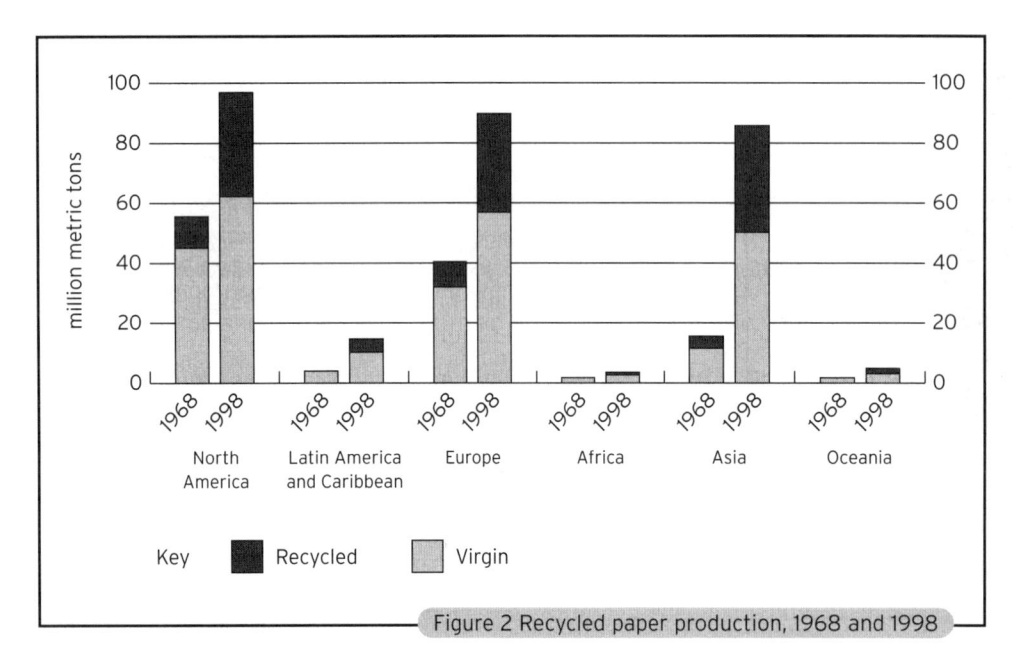

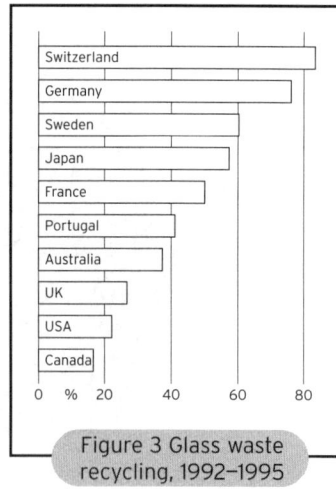

Figure 2 Recycled paper production, 1968 and 1998

Figure 3 Glass waste recycling, 1992–1995

- Increasing the spatial coverage of recycling points for glass etc. At present long trips to the recycling centres can cost more than recycling saves. In Cologne there are over 900 street-side recycling points that are emptied daily.
- Transport initiatives, e.g.:
  - the Congestion Charge in London
  - campaigns to share cars when commuting or even to ban cars with one driver from the outside lanes of freeways as in California
  - car park charging to deter commuting
  - encouraging the use of public transport, e.g. the free passes for the over 60s in London; children travel free on public transport during the school holidays in Austria.
  - encouraging cycling with cycleways and the use of cyclist-friendly traffic lights as in the Netherlands.

QUICK CHECK

1 Does the map in Figure 1 explain why the USA (President Bush) has failed to ratify the Kyoto protocol?

2 What are the merits of local initiatives? What issues do they bring to mind?

2.4 Wilderness environments

Key questions for this section:
- What are wilderness environments?
- Where are they located?
- What are the characteristics of wilderness landscapes and environments?

You will be revising:
- Types of wilderness environments.

This is an excellent option choice because it overlaps with the 'Biomes and ecosystems' sections of Global Challenge (see 1.5–1.9, pages 24–38). It has links to climate, population and aspects of development in the economic section (see 1.13–1.15, pages 52–65). The list below gives you plenty of case study areas. You need to know the physical characteristics of two or three of the groups below (see 1.5, page 24).

Classifying wilderness environments

The list below is based on the definitions of Conservation International.

		Generalisation 1	Generalisation 2	Generalisation 3	Generalisation 4
1	**Tropical rainforests**				
1.1	Amazonia – Brazil, Venezuela, Colombia, Bolivia, Peru				
1.2	Congo forests of central Africa				
1.3	New Guinea				
2	**Tropical woodlands, savannas and grasslands**				
2.1	The Chaco – Argentina, Paraguay, Brazil and Bolivia				
2.2	The Caatinga – Brazil				
2.3	Miombo-Mopane Woodlands and Grasslands – Angola, Zambia, Malawi, Tanzania				
2.4	The Serengeti – Tanzania, Kenya				
2.5	Cape York – Australia				
2.6	Arnhem Land – Australia				
2.7	Kimberley – Australia				
3	**Wetlands**				
3.1	The Pantanal – Brazil, Paraguay, Bolivia				
3.2	The Llanos – Colombia, Venezuela				
3.3	Banados del Este – Uruguay				
3.4	Sunderabans – Bangladesh–India (the smallest wilderness 10 000km²)				
3.5	The Sudd – Sudan				
3.6	The Okovango – Botswana, Namibia				

2.4 Wilderness environments

4	**Deserts**				
4.1	Sonoran and Baja California – USA				
4.2	Greater Chihuahuan – Mexico, USA				
4.3	Mohave – USA				
4.4	Colorado Plateau – USA				
4.5	Coastal deserts of Peru and Chile				
4.6	Sahara and Sahel – Algeria, Libya, Egypt, Sudan, Eritrea, Chad, Niger, Mali, Mauritania, Western Sahara, Morocco, Burkina Faso				
4.7	Kalahari – Namibia, Botswana, South Africa				
4.8	Namib – Angola, Namibia				
4.9	Arabian Deserts – Saudi Arabia, Oman, Yemen, Israel, Egypt, UAE, Bahrain, Kuwait, Iraq				
4.10	Asian deserts – Iran, China, Mongolia, Kazakhstan, Uzbekistan, Turkmenistan, Afghanistan, Pakistan				
4.11	Australian – Australia				
5	**Temperate forests**				
5.1	Northern Rocky Mountains – Canada, USA				
5.2	Pacific NW – USA, Canada				
5.3	Appalachians – USA, Canada				
5.4	Magellanic Subpolar Rainforest – Chile, Argentina				
5.5	European mountains – Spain, France, Switzerland, Austria, Slovakia, Romania, Poland, Italy				
5.6					
6	**High latitude wildernesses**				
6.1	Boreal forests – Canada, USA, Russian Federation, Finland, Sweden, Norway				
6.2	Patagonia – Argentina, Chile				
6.3	The Arctic Tundra – USA, Canada, Greenland, Iceland, Norway, Sweden, Finland, Russian Federation				
6.4	The Antarctic (the largest wilderness 13.9 million km^2)				

Table 1 Wilderness environments appropriate for key questions (shaded)

Defining wildernesses

1 Size – needs to be a distinct biogeographic unit, minimum area 10 000km^2
2 Intactness – natural vegetation prevalent over 70 per cent of area
3 Population density – <5 persons/km^2 (some are higher); 19 of the 38 wildernesses have <1 person/km^2
4 Biodiversity – distinguishes wildernesses from most other areas.

Not all wilderness environments are protected by national legislation; Antarctica is protected by the Antarctic Treaty. Not all are contiguous, e.g. the European mountains, the Boreal forests and the Arctic Tundra.

The US 1964 Wilderness Act defines wilderness as *'an area where the Earth and its community life are untrammelled by man (people), where man himself (the human race) is a visitor.'*

2.4a The significance of wilderness regions

Key questions for this generalisation:

- What is the significance of wilderness regions?
- To what extent do these areas present features that are worth protecting?

You will be revising:

- The protection of wildernesses
- Threats to wildernesses.

You will research the key issue, that of the destruction of wildernesses to satisfy the forces of economic development. You will also recognise that most wildernesses do have an indigenous population. You can use your Tropical Rainforest knowledge from Global Challenge but you should also have contrasting examples.

Why protect wildernesses?

There are numerous reasons why it is important to protect wildernesses. Here are some of them:

- They are the storehouses of biodiversity. Papua New Guinea has over 10 000 endemic species. Rainforests have the richest biodiversity followed by tropical woodlands and grasslands. Obviously biodiversity declines in the high latitude areas with less favourable growing conditions yet there are about 1000 endemic species in the European mountains. High biodiversity is valuable especially in the **hot spots** – the areas of greatest endemism, i.e. where particular species are endemic or confined to that area.
- They maintain ecosystem services such as the provision of clean water, or the retention of vast supplies of fresh water in Antarctica. Ecosystems also fix nitrogen and are home to bees.
- They are carbon sinks, which, if destroyed, would have an effect on climate because of the greenhouse gases released.
- They are areas where indigenous tribal peoples can maintain traditional lifestyles.
- They have a recreational value as people seek out a wilderness experience.
- Their landforms are of global importance, e.g. Antarctica, the Tundra and the savanna plains of the Serengeti.
- They have **aesthetic and spiritual values**. People want to return to convene with nature, and even go back to their roots as expressed by the indigenous cultures and the alternative lifestyles that can be found in these areas.
- **Religious values** have come from wilderness areas. Judaism, Christianity and Islam all have roots in wilderness areas. All link nature to God.
- There is a **moral imperative** to sustain our planet for future generations. The areas listed comprise half the Earth's land surface yet only have 2.4 per cent of the population so they ought to be sustainable.
- Each ecosystem and its products are of value, especially in the case of forests.

Indigenous populations

You may take examples from any of the wildernesses that you have studied. Antarctica is the only area without an indigenous culture. Here, all human life is confined to research stations established by MEDCs.

> **Kimberley Western Australia** (Cl.2.7)
>
> It has a longstanding Aboriginal culture that is almost timeless (the Aboriginal concept of dreamtime) which has been eroded by expropriation of lands for ranching, the establishment of missions, pearl fishing and horticulture. There is now an attempt to re-establish aboriginal control over marginal lands recognising the management abilities of the indigenous population in such a harsh environment.

2.4a The significance of wilderness regions

> **Papua New Guinea** (Cl.1.3)
>
> It has 1100 languages among 6 million people. There is **language diversity** to match biodiversity. Language groups are declining. In most wildernesses languages have died.
>
> Tribal cultures here are still relatively intact and not declining – 97 per cent of the land is still controlled by traditional landowners. Most tribal groups were not discovered until 1930s.
>
> Huli, Asaro (mudmen), Kukukuku (warrior tribe) all have employed traditional slash and burn forest clearing for their agriculture. The Asmat are a swamp dwelling tribe specialising in hunting. Indigenous populations who form 30 per cent of the island's population live at densities of below 5/km^2 (not including the major towns such as Port Moresby).

Threats to wildernesses

Threats to tropical rainforests are similar whatever case study you use. They can be short or long term. The examples here are all from Papua New Guinea.

Short-term threats to rainforest wildernesses in Papua New Guinea include:

- Industrial logging; mobile saw-mills; work of South Korean logging companies
- Unplanned development
- Mining – oil and gas exploitation affecting river systems
- Over fishing and the use of dynamite and cyanide fishing; agricultural clearance leads to soil erosion and deposition in mangrove and sea grass beds destroying fish spawning areas
- Fire
- Roads to open up area; South Koreans granted built roads in return for right to exploit forest 5km either side of the road.

Long-term threats to rainforest wildernesses in Papua New Guinea include:

- Population growth
- Conversion to monoculture, e.g. oil palm plantations, which often follows on from logging; this is a greater threat because it destroys all forest species
- Climate change can lead to fires as in 1997–1998 El Niño year
- Large-scale mega-projects – Ok Tedi mine has altered the river systems in Papua
- Wildlife hunting for live animal trade
- Introduction of species not native to the area.

The key to the significance and value of wildernesses is **fragility;** the ecosystem is in equilibrium, but changes to the inputs will have the effect of destroying the system over a period of time.

This fragility is threatened by:

- Economic development
- Demands for the products of the system
- Expanding populations
- Climate change
- Westernisation
- Imported diseases affecting both humans and the ecosystem itself.

> **QUICK CHECK** Can you produce a similar list of threats to the one above with examples for another contrasting wilderness? Antarctica would be a good case.

2.4b Pressures on wilderness environments

Key questions for this generalisation:

- What are the pressures on wilderness environments?
- How and why are pressures on such areas increasing?

You will be revising:

- Pressures on wildernesses.

These key questions focus on the changes that are taking place in wilderness areas. Again much will depend on the case studies that you research although the principles set out below should apply to most wildernesses.

What are the pressures on wildernesses?

Wildernesses are becoming more **accessible,** not just for tourism, which is geared to satisfy a demand for pristine nature, but also for new groups who exploit people's interest in wildernesses to gain finance to further their own interests, e.g. The Field Studies Council which runs tours in Namibia. It can also be accessibility for exploitation of the ecosystem itself and for the minerals lying beneath it or the waters that flow though it.

Improved accessibility has been aided by **technological developments** such as:

- New aircraft (in 2003 Emirates flew from the UK to Australia with only one stop instead of two cutting two hours off the journey time)
- New cruise ships able to withstand the harsh conditions of the Southern Oceans and Antarctica
- Improved exploration and mining technologies
- Bulk transport technologies such as pipelines and rail systems
- Satellites with Global Positioning Systems (GPS) technology enabling people to 'go it alone'
- Protection against environmental hazards such as extreme cold or heat, and more sophisticated physical protection of mining sites and tourist camps.

Resource demand is rising. At the same time resources in the easily exploited locations are running out. Searches aided by technology are shifting to the unexplored wildernesses, e.g. oil found in Saharan and Sahelian Chad (Cl4.6); timber for plywood used in the construction industries of Japan and South Korea has been the main reason for forest destruction in Indonesia.

The key to the increase in interest in wildernesses is the increasing affluence of the MEDCs.

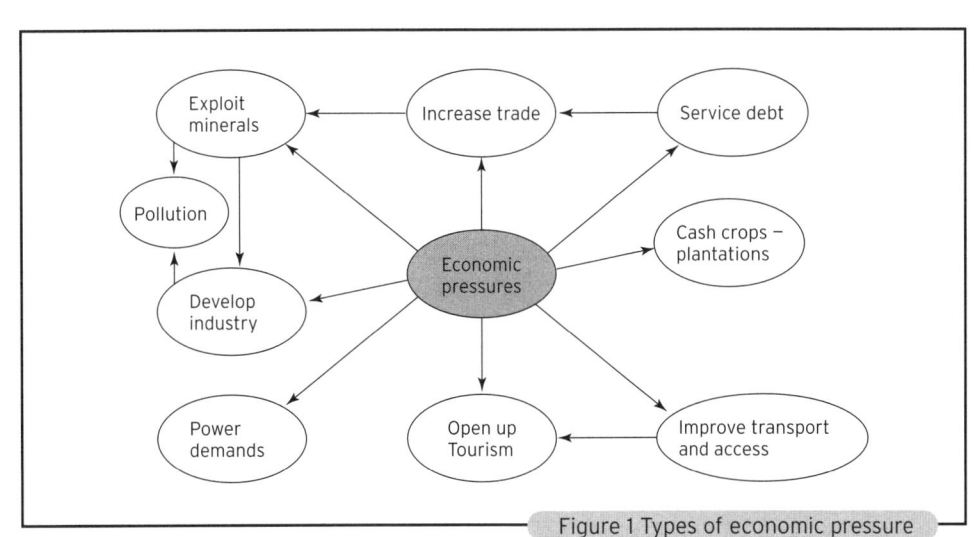

Figure 1 Types of economic pressure

2.4b Pressures on wilderness environments

Table 1 is a conflict matrix for Arnhem Land, Northern Territories, Australia. It is based on the threats of 1) mining of uranium in Kakadu at Jabiru, bauxite at Gove and 2) the existing uses such as ranching aboriginal lands. The area has only 15 000 people. Excluding two small towns the population density is 0.08person/km^2.

	1	2	3	4	5	6	7	8	9	10
1 Environmental destruction										
2 Loss of biodiversity	+									
3 Mining companies, TNCs	X Have to build mines	X Fauna and flora destroyed								
4 Ranchers	? If overgrazed	X If using alien species	X Take lands							
5 Environmental Pressure Groups	+ In favour	+ In favour	X	Neutral						
6 Government	+ Wants exports	+ Wants exports	+ Wants exports	X Handing land rights back	X Tree huggers!					
7 Indigenous people – 30 tribal groups	+ Practices in harmony with environment	+ Harmonious practices	X Opposed to land take over	Over land rights	+ Helps them but like privacy	+ Have support, not always the case				
8 Settlers in area	Neutral	Neutral	+ Want jobs	+ Seen as job provider	X	X No interference	Neutral			
9 Transport developments	X New roads and road trains	X New roads used by road trains	+ Essential for progress	X Slice through lands	X A nuisance	X Invade lands	+	+		
10 Fire regime	Altered	Altered	+	X Might spread to stockland	+ Support practice	X Lost skills of fire break control	+	X	X	

Table 1 Conflict Matrix for Arnhem Land

Key: X = Conflict, + = No conflict

QUICK CHECK

1. Classify the pressures on wildernesses as either long or short term with reference to one wilderness you have studied.
2. How does tourism fit into Table 1? Add a further line to show how it might conflict. You could also add the details to support each conflict. Now draw similar matrices for two different wildernesses.

2.4c Strategies for managing wilderness regions

Key questions for this generalisation:

- How and why might protection constrain or conflict with economic development?
- How can such pressures be managed?
- By whom and with what effects?

You will be revising:

- Managing wilderness regions
- Strategies for wilderness regions.

The key questions focus on the management of the conflicts in 2.4b (see page 14). Much will depend on the case studies that you use but the general principles outlined here should help you structure your research.

Managing wilderness regions

Wilderness protection began in the USA in 1892 with the foundation of the Sierra Club which campaigned for National Parks.

International organisations

- **The United Nations Environment Programme (UNEP)** was founded to 'provide leadership and encourage partnership in caring for the environment by inspiring, informing, and enabling nations and peoples to improve their quality of life without compromising that of future generations'. From 1972 it has been responsible for a range of global strategies that impinge on wilderness management including environmental assessment and biodiversity. Since 1992, UNEP has worked on an **Arctic programme** covering activities developed to compile information for the Arctic region. Much of this is designed to support the Arctic Environmental Protection Strategy process, in particular the **Conservation of Arctic Flora and Fauna (CAFF)** programme. It works with other organisations such as WWF who funded the first circumpolar assessment of the impact of climate change on Arctic breeding water birds. The 2002 International Year of the Mountains led to a UNEP Summit in Kyrgyzstan.
- **GLOBIO** is an organisation focusing on the cumulative impacts of increasing resource demands on man and the environment, in support of UNEP's mandate of harmonising *global environmental assessments*. The GLOBIO project is a partnership between a network of organisations coordinated through the **Norwegian Institute for Nature Research** and *UNEP*.
- **World Wide Fund for Nature (WWF)** campaigns for wilderness protection. Since 1998 it has focused on 200 Global Ecoregions such as the Sudd (see CI 3.5, page 109) and the Namib desert (see CI 4.8 page 110).
- **Greenpeace** campaigns against practices that are harmful to the environment whether in or beyond wildernesses. In 2003 it campaigned against Finnish forest clearance practices.
- **Friends of the Earth** has an international presence in 68 countries and, in particular, fighting TNCs.
- **Conservation International** is a US-based organisation that has campaigned directly for wilderness areas.

2.4c Strategies for managing wilderness regions

National organisations

Much depends on which wildernesses you are studying. If you are studying Kakadu then it might be worthwhile looking at Australian organisations with their strong support. The Wilderness Society was founded in 1976 to protect parts of Tasmania and now runs a campaign for 'Wild Country' and wildlife corridors. Because there are no real wilderness areas in the UK it is not advisable to use UK examples. However, I know that your examiners are quite liberal and would accept the discussion of the National Parks and the Countryside Commission.

Strategies

1. **Ecotourism** is much trumpeted as a solution. However, many areas outside wildernesses use the label to try and attract a certain type of more discerning tourist. So make certain that your case study deals with a wilderness resort. Here are some eco-resorts which you can find on the web: Chumbe Island, Zanzibar; Black Sheep Inn, Ecuador; Daintree Ecolodge, Australia; and Forest House Lodge, Canada.

2. **National Parks** are another strategy that aims at conservation. The largest is the NE Greenland National Park. National Parks are a common governmental strategy which depends on good management for success. Successful examples include the US deserts (CI 4.1, 4.3 and 4.4 Wildernesses). More management problems arise in the African parks such as Serengeti (CI 2.4) and Kruger (CI 2.3) where poaching, overgrazing and hunting have had an effect. While hunting is banned, bush tours and night tours leave trails of vehicle impact on the landscape.

3. **Designation of World Heritage Sites** might be effective for small areas or parts of a wilderness such as the Grand Canyon. It has been proposed for Antarctica but who would manage a continent?

4. **Global Treaties,** such as the Antarctic Treaty 1961, determined the peaceful use of Antarctica. Nevertheless, the management of the increasing tourist visitor count is difficult when no nation manages the area. Even this treaty is under threat as natural resources become increasingly scarce.

Where has management been less successful?

1. In countries without the resources to withstand pressures, e.g. Brazil and Indonesia where the lure of exports has resulted in wholesale clearance of forests (CI 1.1 and 1.3).

2. In areas where the very environment of the wilderness is based on the deprivations of nature, e.g. the Sahel (CI 4.6).

3. Wildernesses close to major populations where the wilderness has become fragmented by development, e.g. European Mountains (CI 5.5).

4. Areas where population pressures have caused environmental degradation, e.g. Himalayas and the Sundarbans (CI 3.4)

5. Areas of poverty such as the Serengeti (CI 2.4) where game wardens have been easily bribed.

6. Areas where defence has taken priority, e.g. Mojave Desert (CI 4.3).

> **Reminder**
>
> Ian Flintoff and Sally Cohen have written *Managing Wilderness Regions*, published by Heinemann 1998, which addresses the topic of strategies for wilderness management especially in Chapter 4.

Quick Check

1. You should look in detail at the work of at least one of the international organisations at the start of this section and assess their contribution to managing the conflicting demands on wildernesses.

2. Use your own case studies to see if they match any of the strategy successes or disappointments listed above.

2.4d Protecting wilderness regions

Key questions for this generalisation:
- Should wilderness regions be protected and conserved?
- What are the ways in which such strategies might be attempted?

You will be revising:
- The political problems of wilderness strategies
- How to evaluate these strategies.

Key questions focus on the central theme of sustainability and an evaluation of strategies. So you need to know the strengths and weaknesses of the strategies.

Evaluating strategies

Figure 1 demonstrates the best way to go about evaluating strategies.

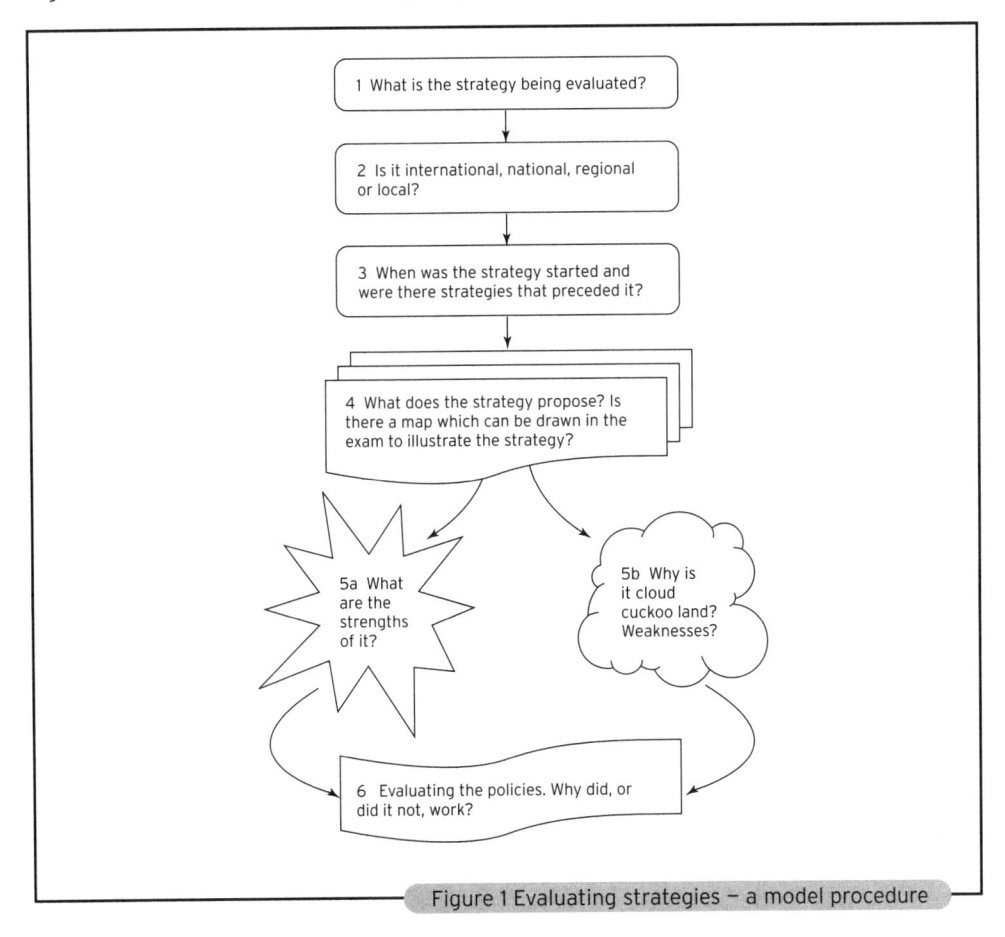

1. What is the strategy being evaluated?
2. Is it international, national, regional or local?
3. When was the strategy started and were there strategies that preceded it?
4. What does the strategy propose? Is there a map which can be drawn in the exam to illustrate the strategy?
5a. What are the strengths of it?
5b. Why is it cloud cuckoo land? Weaknesses?
6. Evaluating the policies. Why did, or did it not, work?

Figure 1 Evaluating strategies – a model procedure

The problem of boundaries

If you look at the list of wildernesses in 2.4 (pages 109–110) it is soon apparent that many straddle national boundaries. Once there are different legal and financial systems, creating a successful strategy may become impossible. The greatest strategy successes have come where the wilderness is within a single country. The Antarctic is the sole case of international agreement and could become a blueprint for the future of other wilderness management strategies.

2.4d Protecting wilderness regions

Antarctica

Antarctica is an extremely fragile ecosystem and the world has agreed to keep it protected. However, people have already made some claims on it:

- There are 36 research stations. It is a continent that is important for scientific study, e.g. climate change and the hole in the ozone layer
- It is known to have oil and minerals below its frozen wastes, e.g. oil in the Ross Sea
- It already has a tourist economy
- Fishing is governed by the Convention of Antarctic Marine Living Resources 1982.

Timeline:

1940	Seven countries laid claim to the continent by 1940.
1945–1960	The military threats were increasingly apparent during the Cold War.
1958	The original **Antarctic Treaty**
1961	**Antarctic Treaty** – did not cover mineral resources.
1972	Greenpeace first proposed a World Park status for the area. Supported since by New Zealand.
1980	**Convention for the Conservation of Marine Living Resources** becomes part of the Treaty.
1991	**Environmental Protocol** banned all mineral activities until 2041. It also bans activities unless a Comprehensive Environmental Evaluation carried out. Flora and fauna received increased protection including the removal of all dogs. Waste was to be disposed of away from the continent and marine discharges stopped. Polluter-pays regulations are being introduced.

There are two levels of signatures to the Antarctic Treaty:
(i) Consultative Parties – 26 countries who are actively involved in research.
(ii) Non-Consultative Parties.

Those countries with a tradition of Antarctic exploration did find the internationalisation difficult at first, e.g. the UK, but it is accepted today.

International progress in the Namibian Desert

Namibia has proposed the **Three Nations Namib Desert Transfrontier Conservation Area,** an international continuum of conserved lands. These stretch from Richtersveld National Park in South Africa through Asis Hot Springs Game Park north through other parks to the Skelton Coast Park and into Angolan Iona National Park. The area includes one of the world's biodiversity hotspots and therefore merits conservation. A treaty was signed between South Africa and Namibia in 2001 as the first stage in the process. When all have signed, 74 per cent of the Namib will be protected.

> **QUICK CHECK**
> 1 Apply the evaluation sequence in Figure 1 to at least two wilderness strategies.
> 2 For any wilderness that you have studied list the strategies used to protect the wilderness. Try to group them as international, national and local.

2.5 Examination questions

Unit 5, Paper 1 (6745/1)

The examination generalisation will be issued by Edexcel about four weeks prior to the examination. The examination will comprise two questions on each option based on the pre-released generalisation question. The questions below are examples related to the generalisations.

Environments and resources

Generalisation 3 (see 2.1, pages 79–80)
1) Is the global energy crisis the same for everybody?

Living with hazardous environments

Generalisation 1 (see 2.2a, pages 84–85)
2) Discuss how far natural hazards can be viewed as purely physical processes.

Generalisation 2 (see 2.2b, pages 86–88)
3) With reference to a range of hazard events, discuss the opinion that the impact of hazards is more a result of their frequency than their magnitude.

Generalisation 3 (see 2.2c, pages 89–90)
4) With reference to a range of hazard types, explain why people react in different ways to similar hazards.

Generalisation 4 (see 2.2d, pages 91–93)
5) Are natural hazards occurring more frequently? Critically assess this statement.

The pollution of natural environments

Generalisation 1 (see 2.3a, pages 96–98)
6) Pollution is an inevitable by-product of development. Critically assess this assertion.

Generalisation 2 (see 2.3b, pages 99–102)
7) Assess the view that the economic costs of pollution are far greater than either the environmental or social costs.

Generalisation 3 (see 2.3c, pages 103–105)
8) To what extent does the management of pollution incidents depend on where they happen and who caused them?

Generalisation 4 (see 2.3d, pages 106–108)
9) Controlling pollution is an international imperative yet is more successful locally. To what extent is this the case?

Wilderness environments

Generalisation 1 (see 2.4a, pages 111–112)
10) Wildernesses are fragile environments. To what extent does this apply to every wilderness?

Generalisation 2 (see 2.4b, pages 113–114)
11) To what extent are the conflicts between resource developers and indigenous peoples irreconcilable?

Generalisation 3 (see 2.4c, pages 115–116)
12) Is it possible to resolve the conflict between development and conservation of wilderness regions?

Generalisation 4 (see 2.4d, pages 117–118)
13) To what extent is it possible to have sustainable use of wilderness areas?

> **Reminder**
>
> Ask your tutor to give you a copy of the generic markscheme.

PART 3
Researching Global Futures: Challenges for human environments

General principles and strategies

Part 3 covers the second paper of the specification Unit 5 (6475/2): **Challenges for human environments**. You will be expected to choose one of the options below:

Development and disparity
Feeding the world's people
Health and welfare
The geography of sport and leisure

This is the second paper of examination 6475/2. It is one where you can shine because it will be all your own work. You are expected to research an essay topic from a choice of 24 topics (six from each: see 3.1–3.4, pages 123–130). It is possible that there may be only five topics per option in the future, i.e. after 2005. Don't spend too much time on the paper because it is only worth 7.5 per cent of the A Level mark. But do spend enough time. The mark is combined with that for paper 6475/1.

Selecting your essay title

It is possible that your school has restricted your choice to just one of the four options. Other schools and colleges may offer a choice across a range of options. Either way you need to make a careful selection.

- The most attractive title may not be the easiest. Spend time on the choice.
- Study the Foundation section before you make any choice.
- Use this revision guide to assist with your choice of topic.
- Discuss your potential title with your teacher who will be able to advise you on your choice. The teacher knows your ability and, just like high diving, some topics are more difficult and complex for some students.
- Always try not to be the only person doing a title unless you are really confident and the teacher can support your efforts. When more than one person does a title there is the opportunity to share the research time and have mini-seminars to develop your ideas.

Focus on the question

- Circle or highlight the **key subject words** such as 'agribusiness', 'food supply', 'bilateral aid', 'social welfare and economic welfare', and 'development and distribution of sports facilities'.
- Circle or highlight the **command words** such as; 'analyse', 'assess', 'compare', 'discuss', 'evaluate', 'examine', 'examine the extent to which', 'explain', 'how far do you agree', and 'to what extent'. All of these commands appeared in the 2004 series of essay titles.
- Are there spatial or area commands such as:
  - 'global scale' – your examples should been drawn from a series of countries; 'national scale' – your examples should be drawn from a single Country; 'regional scale' – your examples should come from a defined area of a country; 'local scale' – examples should be from a named small area such as your home town.
  - 'country' – a country as recognised either by the UN or one of the countries of the UK.
  - 'large urban area' – normally a city of 100 000+ but those of you living in more rural areas could select towns above 50 000.

> **Reminder**
>
> Can you remember what these commands mean? Check on page 132 to see if you have understood the commands.

3.1 Researching Global Futures: Challenges for human environments

- 'LDC, LEDC, NIC, MEDC' – check these terms.
- 'named area' – name the area, e.g. South Wales, and stick to it unless a contrast is demanded.
- 'two contrasting countries' – it is preferable to use one LEDC and one MEDC rather than two LEDCs.
- 'The North' and 'The South' – the Brandt terms for areas of contrasting global development and not the areas of the UK!

Starting research; planning research

1. Read the basics in a textbook or in the relevant section of this guide.
2. Read any other articles suggested by the teacher.
3. Read the more advanced, specific texts suggested by your reading.
4. Start website searches.
5. Always record what you found where so that you can provide good references in your bibliography.

Progress reports

1. Always have a timetable of key progression points and keep to it.
2. Arrange or participate in arranged seminars that focus on your option. It is good to listen to the plans of others who might not be doing your option. You will learn how others are organising their work.
3. Begin to think about constructing your own diagrams and tables.
4. Put your research onto summary cards. Alternatively keep an organised research file including press cuttings. Share resources with others doing the same title.
5. Start to develop a plan for the essay.
6. Keep a running bibliography with comments about bias and purpose of article noted.

Writing up

1. The word limit is 1500 words. There is a 10 per cent margin so 1700 words is just about tolerable. A report is more concise than an essay.
2. Do NOT try to get away with writing too little.
3. Diagrams, maps and Information Boxes (Text boxes) do not count. Abstracts and bibliographies do not count. A picture (map/diagram) is worth a 1000 words. Case studies should always support the case being made.
4. Appendices are not the best way to cram extra in. Box it, because it is then part of the main text.
5. Make sure that Figures, Tables and Text boxes are all numbered separately. If you select to use a report style, you can number the paragraphs separately. Numbering that is continuous between text and boxes is incorrect.
6. The best way of referencing is called the Harvard system (Author/date) in text with the full reference in the bibliography Author/Date/Book or Article Title, Publisher. You can use footnotes if you prefer.
7. The sequence and suggested words are as follows:
   - Contents, list of figures and text boxes
   - Abstract – a summary of the whole (NOT I did this and then that – see websites or journals for examples of Abstract style if you are not sure what to write here)
   - Introduction and definition of terms (150 words)
   - The main body of the text (1200 words). This should contain your research findings and an interpretation of your research in the light of the question.
   - Conclusion and evaluation of the results (150 words)
   - Bibliography including any bias comments especially for websites.
8. There is no need for a Methodology section unless you want to put it in a box.
9. Remember to keep organised – organisation and time budgeting are keys to success.

10. There are 10 marks for the quality of your written communication. Poor English, text message English, weak grammar, incorrect punctuation or spelling, especially of geographical terms, will all lose you marks. QWC should be faultless.
11. Do not plagiarise, i.e. copy material direct from essays on the web or other documents without acknowledging your source. Plagiarism is subject to heavy penalties.

Getting top marks

The paper is marked according to the same set of criteria that are used for 6475/1.

To score well you need to make sure that the sections of your report have the following qualities:

- **Introduction** You will clearly state what the issue is. You will briefly define your terms. You will introduce the locational context of your work.
- **Research** You will show in the text and in a good bibliography that you have read and researched widely and that your examples are appropriate for your theme. Maps and diagrams will be integrated and used effectively to support what you say. Your bibliography might wish to draw attention to the bias of your sources.
- **Understanding** You will have organised your material in a logical, well-argued and analytical way that applies it to the question. You will recognise that there is a range of perspectives on any issue. Case studies may be used in boxes and the text comments on these and shows how they make a particular point.
- **Conclusion** The report may contain ongoing evaluation and mini-conclusions as it progresses. The final section/paragraph is a summary of the study and its conclusions, which looks back to the introduction to demonstrate how the topic has been covered. (Some people write the introduction last to make certain that it links to the conclusion.)
- **Quality of Written Communication** It goes without saying that it will be a coherent report that keeps to the question. It will use the right terminology from geography and other disciplines. There will be no spelling errors (use spell checkers) and paragraphing will make it a joy to read.

> **Reminder**
>
> Ask your tutor to give you a copy of the generic markscheme

Remember

- Make the study user-friendly for the reader/examiner.
- It is a research project/essay and not just a repeat of a textbook.
- Use figures to summarise data and trends. Cross refer to them in the text.
- Place focused case studies in the text boxes and use the text to comment on the examples.
- Check that you have answered the question.
- Use the spell checker but get a friend to read it to check on slips such as There/their or were/where which spell checkers miss.
- You may use bullet points.

In the four sections that follow there is a guide to some of the reading and websites that you can use to assist your research. These listings are not exhaustive and only provide you with some starting points. Much will depend on the essay titles set each year. In addition the Edexcel website provides guidance.

3.1 Development and disparity

You will be researching the causes and consequences of uneven development and its effects on people and environments. You have the chance to research the effects of development schemes and projects and be able to critically reflect on these attempts to provide economic and social development. This option links with Sections 1.13–1.17 (pages 52–70) in this guide (Enquiry questions 4.10–4.12 on the Global Challenge paper 6474). The sources here are just a few to get you started. Your teacher should know of others and have updated lists. You should begin any research by consulting the books, journals and websites in the Foundation section before proceeding to your essay's enquiry question.

Research

Foundation question

How can development and disparity be measured? How useful are different criteria in measuring development?

Books and reports:
B. Digby and N. Yates, *Global and Regional Disparities*, Heinemann 1997
R Hodder, *Development Geography*, Routledge 2000. Good coverage of issues and excellent bibliography.
A. Read, *Inequality and development*, Bell and Hyman 1985. An introduction to the topic.
R. Robinson, *Alternative Approaches to Development*, in B. Digby, Global Futures Options, Heinemann, 1997
M. Todaro, *Economic Development*, Longman 1998.
UNDP, *Human Development Report*. An annual publication. Good unbiased statistics.
T. Unwin, *Atlas of World Development*, Wiley, 1994.
World Bank, *World Development Indicators 2001*. Good statistics.
World Bank, *World Development Report, Knowledge for Development*, OUP 1999.

Magazines:
New Internationalist – a left wing publication which makes you think.
Third World Planning Review
The Ecologist

> **Web link**
> For more research, go to the following website and enter the express code 1552S:
> www.heinemann.co.uk/hotlinks

Enquiry questions

1 What is the Development Gap? At what scale does it exist and why is it so significant?

Books and reports:
W. Brandt, *North–South: a programme for survival*, Pan, 1980
A. Crump, *A–Z of World Development*, New Internationalist, 1999
R Prescott-Allen, *The Wellbeing of Nations*, Island Press, 2001
The World Guide, New Internationalist, annual. A good summary of developments by country – latest is 2003–2004.

> **Web link**
> For more research, go to the following website and enter the express code 1552S:
> www.heinemann.co.uk/hotlinks

2 What are the consequences of disparity, and their continued existence?

Books and reports:
GeoFactsheet No 78 1999 About debt
'The Debt Burden', *Understanding Global Issues No. 93*
S. Mitchell, *Inequality and Poverty – a spiral of despair* Jubilee Research 2003 (on website)
Fr. Felix Raj, 'Global Disparities', *The New Statesman*, March 2002
J. Rigg, *Southeast Asia*, Routledge, 1997, Chapters 3 and 4
J. Seabrook, 'World Poverty', *New Internationalist* 2003

3 How do sources of investment and aid vary? What are there impacts? Can investment and aid affect disparity and development?

Books and reports:
G. Cho, *Global Interdependence Trade, Aid and Technology Transfer*, Routledge, 1998
IIED, *Unlocking Trade Opportunities*, 1998

4 What alternatives strategies exist for future development? How can benefits for minority or less disadvantaged groups be provided?

Books and reports:
R. Greenhill and E. Sisti, *Real Progress Report on HIPC*, Research Jubilee 2000, 2003. Calls for reform of the HIPC programme
IIED, *Sustainable Trade – who benefits?* 1999
G. Monbiot, *The Age of Consent: a manifesto for a new world order*, Flamingo, 2003
G. Monbiot, 'How to Stop America', *New Statesman*, 9 June 2003
D. Ransom, 'Fair Trade', *New Internationalist*, 2001
'The Road from Rio', *New Scientist*, 17 August 2002
World Bank's *Heavily Indebted Poor Countries* 1996 (HIPC) An initiative which began with Mauritius, Nicaragua and Tanzania in 1999.

> **Web link**
>
> For more research, go to the following website and enter the express code 1552S:
> *www.heinemann.co.uk/hotlinks*

Models of development

You should be prepared to demonstrate knowledge of the following models;

- Rostow's Stages of Economic Growth model – only as the most basic starting point
- Friedmann's Core Periphery model
- Myrdal's Cumulative Causation model
- Bale's Cycle of industrialisation model
- Nurske's Vicious Circle of Poverty theory
- Frank's Dependency theory
- Marx's Political model of development.

3.2 Feeding the world's people

This option is concerned with the outcomes of the global population explosion and the ability of the world to feed itself. It provides the opportunity to investigate the effects of food demand on the environment. It also enables you to research the effects of technological approaches to food supply and their impact. It includes marine food supplies. This option ties up neatly with Section 1.11 (pages 43–46) and Sections 1.13–15 (pages 52–65) in this guide (Enquiry questions 4.8 and 4.10–4.12 of Global Challenge 6474). The sources here are just a few to get you started. Your teacher should know of others and have updated lists. You should begin any research by consulting the books, journals and websites in the Foundation section before proceeding to your essay's enquiry question.

Research

Foundation question

What are the current issues in feeding people? Where are these issues most apparent?

Books and reports:
R. Dickenson, *The Geography of the Third World*, Routledge, 1996
E. Millstone and T. Lang, *Atlas of Food*, Earthscan
D. Mitchell, *The World Food Outlook*, Cambridge, 1998
M. Raw and P. Atkins, *Agriculture and Food*, Collins, 1999
G. Tansey, *The Food System*, Earthscan, 1999
M. Witherick, *Food, Farming and Famine*, Stanley Thornes, 2001
FAO, *The State of Food and Agriculture*, UN Annual Report 2002 (on food insecurity)
M. Witherick, *Development, disparity and dependence*, Stanley Thornes, 1998

Journals:
New Scientist, The Ecologist, The Farmers Weekly, ITDG Annual Reviews, The Grocer, New Internationalist.

> **Web link**
>
> For more research, go to the following website and enter the express code 1552S:
> www.heinemann.co.uk/hotlinks

Enquiry questions

1 How and why does food supply vary spatially? How have areas of surplus and famine emerged?

Useful references:
T. Allen, *Poverty and Development in the 21st Century*, OUP, 2000
I. Bowler, *Agriculture in Developed Market Economies*, Longman, 1996
I. Bowler, *Agricultural Change in Developed Countries*, Longman, 1997
B. Ilbery, 'Changing geographies of global food production', Chapter 9 in P. Daniels et al (Eds.) *Human Geography*, Prentice Hall, 2001
A. Macmillan, 'Famine, The Unnatural Disaster', *Geography Review* 1991 No 1
T. R. Reid, 'Feeding the Planet', *National Geographic*, October 1998
G. M. Robinson, 'The World Food Problem', *Geography Review* Vol. 6 No 4 1993. A useful introduction
J. Seager, *The State of the Environment Atlas*, Penguin 1995, especially maps 1–5, 20–21 and 25–35.

2 What developments have attempted to increase the global food supply? By whom have these been carried out?

Books and reports:
L. Anderson, *Genetic Engineering, Food and Our Environment*, Green Books, 1999
G. Conway, *The Doubly Green Revolution: Food for All in the Twenty-First Century*, Cornell University Press, 2002
J. McNeill, *Something New under the Sun*, Allen Lane, 2001. Chapters 5 and 6 good on evolution of water control, dams and irrigation
K. Murwira, *Beating Hunger; The Chivi Experience*, Intermediate Technology Productions, 2000
N. Parrott and T. Marsden, *The Real Green Revolution*, Greenpeace 2002

Websites:
For more information, see the web link in the margin.

> **Web link**
>
> For more research, go to the following website and enter the express code 1552S: *www.heinemann.co.uk/hotlinks*

3 What impacts have resulted from attempts to increase global food supply? Who has been most affected and why?

Books and reports:
J. Kimbrell, *Fatal Harvest: the Tragedy of Industrial Agriculture*, Island Press, 2002
'Politics of food and farming', *New Internationalist* Jan/Feb 2003
'World Fishing, beyond Sustainability', *Understanding Global Issues*, May 2002
Action Aid Pack on Ethiopian food issues. 2002
'World Fishing', *Understanding Global Issues,* No 8/1994

4 What strategies might be attempted in the future to resolve issues in food supply? To what extent can these strategies resolve issues of surplus and famine?

Books and reports:
Norberg, Hodge, Merrifield and Gorelick, *Bringing the Food Economy Home: Local alternatives to global agribusiness*, Zed, 2003
'The Future of Farming', *Understanding Global Issues*, No 7/1997,

Key theorists

Thomas Malthus, Esther Boserup, Club of Rome, Karl Marx

3.3 Health and welfare

Health and welfare is a key challenge for this century. We need to know how diseases spread and what their effects are on people, the economy and the environment. Geographers are interested in the environmental causes of some diseases. You will be researching why there are such great variations in life expectancy both globally and locally. There are so many current issues that make this a valuable research topic. It all relates to Sections 1.10–1.11 (pages 39–46) in this guide (Enquiry questions 4.7–4.9 in Global Challenge paper 6474). The sources here are just a few to get you started. Your teacher should know of others and have updated lists. You should begin any research by consulting the books, journals and websites in the Foundation section before proceeding to your essay's enquiry question.

Research

Foundation question

What do we mean by a geography of 'health and welfare' How might such a geography be measured?

Books and reports:
B. Digby, *Health and Welfare*, Heinemann, 1997
B. Digby, *Geography of Health*, Longman, 1995
S. Curtis and A. Taket, *Health and Societies; changing perspectives*, Arnold, 1995
A. C. Gatrell, *Geographies of Health*, Blackwell, 2002
K. Jones and G. Moon, *Health, Disease and Society*, Routledge, 1991 A valuable introduction
J. Lloyd, *Health and Welfare*, Hodder and Stoughton, 2002 The starting text
M. Meade and R. Erickson, *Medical Geography*, Guildford Press, 2000.
M. Witherick, *States of Health and Welfare*, Stanley Thornes, 2002

Journals:
British Medical Journal, New Scientist, The Economist, Journal of Epidemiology and Community Health, The Lancet

> **Web link**
>
> For more research, go to the following website and enter the express code 1552S:
> www.heinemann.co.uk/hotlinks

Enquiry questions

1 How is geography linked to studies of health and welfare? Why are there geographical variations in healthcare and welfare?

Books and reports:
A. Cliff, *Deciphering Global Epidemics*, CUP, 1997
A. Cliff, P. Haggett, and M. Smallman-Raynor, *Atlas of Disease Distributions*, Blackwell, 1992
J. Dorling, *A New Social Atlas of Britain*, Wiley, 1995; Section 5 Health – mainly based on 1991 census but useful introduction to spatial variations.
Shaw and Dorling, *Health, Place and Society*, Prentice Hall, 2002.

2 Why do spatial variations in patterns of health and welfare exist?

Books and reports:
P. Alcock, *Understanding Poverty*, Macmillan, 1997
P. Elliot, et al, *Spatial Epidemiology*, OUP, 2000.
R. Mansell Prothero, 'Health Hazards and Wetness in Tropical Africa', *Geography* Vol. 85 Pt. 4 2000
P. Townsend, *Inequalities in Health; the Black Report*, Pelican, 1992

3 What impacts does disease have on societies, the economy and the environment? How do such impacts arise?

Books and reports:

T. Barnett and P. Blackie, *AIDS in Africa*, Belhaven, 1992

P. Gould, *The Slow Plague; a Geography of the AIDS pandemic*, Blackwell, 1992

S. Usdin, 'No nonsense guide, HIV/AIDS', *New Internationalist*, 2003

'Atishoo, Atishoo, we all fall down', *New Scientist* 24 November 2001 Looks at Black Death

4 What challenges are there for the future in health and welfare? To what extent can these challenges be met?

Issues such as ageing, new diseases such as SARS and chicken flu and the costs of health and welfare care are very dependent on reading the quality press.

Books and reports:

R. Kitchen, *Disability, Space and Society*, The Geographical Association, 2000J.
Mohan, *A National Health Service*, Longman 1995

R. Wilkinson, *Unhealthy Societies*, Routledge, 1996

'Caring for the Citizen', *Understanding Global Issues*, No 96

'The Conquest of Disease', *Understanding Global Issues*, No 98

B. Crumley, 'Elder Careless', *Time Magazine,* 1 September 2003. Impact of August heat on death rates in France.

S. Jackson, 'Age Concern? The geography of a greying Europe', *Geography*, Vol. 85 Pt. 4 2000

> **Web link**
>
> For more research, go to the following website and enter the express code 1552S:
> *www.heinemann.co.uk/hotlinks*

Key models and theories

Demographic transition
Epidemiological transition
Hagerstand's Diffusion theory
Multiple causality – see Jones and Moon
Infectious Disease Model – see Jones and Moon
Chronic Disease Model – see Jones and Moon

3.4 The geography of sport and leisure

Leisure activities are one of the fastest growing fields of employment. The expansion of sport and leisure, and the prominence that they are given, create research opportunities. You are able to study its increasing globalisation, the impact of sports and leisure facilities and the management of developments. There are some links to 'Globalisation' in Section 1.14 (pages 58–63) and 'Conservation' Section 1.9 (pages 36–38 in this guide). The sources here are just a few to get you started. Your teacher should know of others and have updated lists. If your college has a sports studies/science department, consult them. You should begin any research by consulting the books, journals and websites in the Foundation section before proceeding to your essay's enquiry question.

Research

Foundation question

What are the spatial aspects of sporting and leisure activities? How do sporting and leisure activities pose questions about the use of space?

- Aspects such as frequency and magnitude of use of facilities and landscapes.
- Distribution of facilities in a country, region, city.
- The question of overuse and its environmental effects.
- Environmental degradation.

Books and reports:
J. Bale, *Sports Geography*, Routledge, 1989; E. Dunning, *Sport Matters*, Routledge, 1999; J. Edmonds, *At Leisure*, Hodder and Stoughton, 1995; S. Glyptis, *Leisure and the Environment*, Belhaven, 1993; R. Prosser, *Leisure and Tourism*, 2000; S. Warn and M. Witherick, *A Geography of Sport and Leisure*, Stanley Thornes, 2003 (Written by the Chief Examiner)

Websites:
Search the web for individual sports. Do not just focus on one sport. Soccer, rugby, cricket, swimming, athletics all have sites worth exploring for detail of history and spread of activities.

For more information, see the web link in the margin.

Journals:
Journal of Leisure Sciences, Journal of Leisure Research, Leisure Management, Tourism Management, The Climber, The Golfer, Ski Monthly

> **Web link**
>
> For more research into sport and leisure go to the following website and enter the express code 1552S: www.heinemann.co.uk/hotlinks

Enquiry questions

1 How does sport vary spatially? Why does it happen?

Some key themes here are:
- The regional origins of sports, e.g. Rugby and the British Empire; and sports stars, e.g. Ethiopian and Kenyan long distance runners.
- The role of the environment for outdoor sports, e.g. the relatively small number of links based golf courses because need sand dunes on the coast. The rise of indoor stadia.
- Government investment e.g. in former Soviet Bloc. Local investment.
- At a local level, planning policies to provide access to facilities.
- The role of levels of development and relative affluence.

Useful references
J. Bale, *Sport, Space and Society*, Blackburn Press, 2001
P. Waylen and A. Snook, 'Patterns of regional success in the football league', *Area*, Vol. 22 Pt.4

2 What impacts do sport and leisure and their development have upon people, the economy and the environment? How far do these impacts bring benefits or problems?

The main themes:
- The costs and benefits of the development of facilities including the development of stadia for major events such as the Olympics.
- Economic multiplier effects.
- Exposure of sports and increased demand for facilities.
- The impact of sponsorship.
- Where do the stars live? Is money returned home?

Books and reports:
J. Bale, *Sportscapes*, The Geographical Association, 2000; J. Coakley, *Sport in Society: Issues and Controversies*, McGraw Hill, 2003; M. Orams, *Marine Tourist Development, Impacts and Management*, Routledge, 2000; 'Globalised Sport; Media Money and Morals', *Understanding Global Issues*, Sept.2000.

3 What are the relationships between the geography of sport and leisure, and the process of economic development?

The main themes include:
- Sportswear makers and economic development.
- Sport and national consciousness in the struggle for development.
- Exclusion of LEDCs from global sporting event venues.
- Leisure developments as inward investment.

Books and reports:
A. Cooke, *The Economics of Leisure and Sport,* Routledge, 1994; C. Gratton, *The Economics of Sport,* Spon, 2000; J. Lea, *Tourism and Development in the Third World*, Routledge, 1988; P. Pattullo, *Last resorts, the costs of tourism in the Caribbean*, Cassell, 1998

4 What alternative models of leisure activity exist for the future? How can these be evaluated? To what extent are they sustainable?

Books and reports:
WTO, *Guidelines for the Development of National Parks and Protected Areas*, WTO and UNEP, 1992

Look up websites for football clubs contemplating relocation and redevelopment: Arsenal and Portsmouth have yet to start; St Mary's Southampton; Walker Stadium Leicester and the City of Manchester Stadium are now completed.

Models of the distribution of leisure facilities in your home area can be developed using the local authority website to locate provision and outline policies.

> **Web link**
>
> For more research into sport and leisure go to the following website and enter the express code 1552S:
> *www.heinemann.co.uk/hotlinks*

Key models and theories

1. Sporting hotspots.
2. Rostow's model of Economic development as applied to sport and leisure.
3. New International Division of Labour applied to the production of sports equipment and sportswear. Could this be applied to the origins of sports stars?
4. Theories of Migration as applied to international sports stars home country/area and their subsequent movement to new clubs or training bases – track the career of the Current Arsenal or Chelsea team and compare it with Kenyan athletes.
5. Globalisation – Real Madrid, Manchester United, Harlem Globetrotters, Formula 1 racing, horse racing for top horses, owners and jockeys.

3.5 Sample essay titles

The essay titles which follow are examples of the types of essay that you may be expected to write. The titles are released in May for the following year's cycle of examinations (i.e. both January and June). They are all on the Edexcel website. You only have to submit ONE essay on ONE of the four options.

Development and disparity (see 3.1, pages 123–124)

1. With reference to either a large urban area or a named rural area, *describe and offer reasons for* the geographical distribution of deprived groups.
2. *To what extent* do models of regional development help us to understand how disparities develop within a country?
3. *Critically evaluate* the range of criteria used by geographers to measure global levels of development.
4. *Discuss* whether bilateral aid is distributed equitably.
5. *To what extent* is the development gap decreasing?
6. Trade Blocs both aid and hinder development. *Discuss.*
7. The Johannesburg Earth Summit set itself the task of combating poverty. *Is this* the primary task of development?
8. With reference to a range of examples, *evaluate* the merits of top-down and bottom-up approaches to development.
9. *To what extent* is aid merely a reflection of the source, nature and expectations of the donors?
10. *To what extent* is trade the most viable international action to solve the development gap?

Feeding the world's people (see 3.2, pages 125–126)

1. Under-nutrition is a developing world issue whereas over-nutrition is a developed world problem. *Discuss.*
2. *Explain why* the crisis predicted by Malthus's model of population growth and food supply has not come about.
3. *To what extent* are low technology and organic farming the only sustainable futures for farming?
4. *Are* agribusinesses creating more problems than they solve?
5. Farming the sea is as threatening to the environment as farming the land. *Discuss.*
6. The technology of the Green Revolution solved food problems for three decades. *How far do you agree* with this statement?
7. Increasing food supply inevitably brings people into conflict with the environment. *Discuss.*
8. Overproduction of food has as many dangerous consequences as underproduction. *Discuss.*
9. Famine can be alleviated by food aid. *Discuss this assertion* with reference to a range of areas suffering from famine.
10. *To what extent* are the global variations in food supply a product of the natural environment?

Heath and welfare (see 3.3, pages 127–128)

1. *Evaluate* the roles of wealth and poverty in determining good health either in a country or in a region.
2. The challenges for health and welfare in the future are markedly different. *Explain why* this is so.
3. With reference to either one animal disease or one human disease, *explain why* its economic and social consequences can vary spatially.
4. *Examine* the role of the physical environment in the incidence and spread of diseases.
5. *Examine* some of the new infections in terms of their environmental origins and their effects on society.
6. *Who should provide* health care and society's welfare in the future?
7. *To what extent* are welfare conditions more spatially confined than those of health?
8. Pollution is the biggest threat to health and welfare. *Discuss.*

9 *Critically evaluate* the indicators that might be used to draw attention to the major areas of health and welfare and where there is a deficiency at either a global or a local scale.

10 *To what extent* does welfare provision differ between urban and rural areas?

The geography of sport and leisure (see 3.4, pages 129–130)

1 Upland environments are too fragile for intensive leisure use. *To what extent* is this true of all upland areas?
2 With reference to a range of events, *assess the extent to which* major sporting events are a focus for inward investment.
3 The 'new sports' have a much greater impact on the environment than the long-standing sports. *Discuss*.
4 *Why and how* do sport hotspots develop?
5 *What are* the environmental issues posed by an ever-increasing participation in either leisure or sporting activities?
6 *To what extent* is the private provision of leisure taking over from the public provision of leisure?
7 The globalisation of sport has caused a complex pattern of international migration. *Explain why* this should be so?
8 The geography of sport and leisure further emphasises the basic division in the world between the 'haves' and the 'have nots'. *Discuss*.
9 *What* conflicts can arise from the designation of space for either sports or leisure facilities?
10 *To what extent* is the involvement of trans-national companies in sport a benefit to the economy and society of countries?

The command words have been *italicised* in all of these questions. The most common commands are:

- *'To what extent'* – you must evaluate several possible explanations and attempt to say which are the most plausible.
- *'Evaluate'* is asking you to do the same.
- *'Critically evaluate'* is similar in that it expects you to assess the reasoning.
- *'Assess the extent to which'* is another evaluative question that expects you to evaluate almost on some linear scale from least to most appropriate.
- *'Discuss'* is asking you to outline and explain a range of factors. Sometimes the question expects some evaluation.
- *'Why and how'*, *'What are'* and *'Explain why'* are all asking for a logical, supported set of reasons.
- There are some questions like 7 (Section 3.1), 4 (Section 3.2) and 9 (Section 3.4), that do not have the normal commands, but all expect you to develop an argument in a logical fashion supported by examples.

PART 4

Synoptic unit: issues analysis

General principles and strategies

The Synoptic paper is Unit 6 (6476) in the specification and, as its name implies, this is the unit in which you are expected to pull together **all** the elements of the course that you have studied over your AS and A2 studies. In many ways it is similar to writing a thesis at degree level.

Prior study points

1. Make sure that you have practised on whole past papers and not just parts of them. You should have attempted at least ONE complete paper as a mock so that you are used to the format.
2. You cannot revise for this paper by just using the resource materials. You have to show that you have completed the whole course. This is why it is called synoptic.
3. Do not expect to be able to predict the questions. Learning for your predictions, or indeed those of your teacher, may mean that you do not answer the questions.
4. The paper does expect you to be able to draw on ALL of your studies where necessary and not just the resources.
5. Remember to keep your AS notes in good order and perhaps consult the AS volume that accompanies this book.
6. Throughout your course you should have developed the ability to read and interpret maps (including Ordnance Survey maps), diagrams, graphs, cartoons and photographs (including oblique, aerial and satellite images).
7. As this is a synoptic paper, markers will credit comments that demonstrate your broader geographical knowledge.
8. Look at some of the issues analysis and decision-making exercises that are in the Heinemann texts relating to this specification and especially those edited by Bob Digby.

When you receive the Advance Information Resource Booklet

The advanced materials are issued at least two working weeks before the examination itself. This is frequently more than two actual weeks and is planned to be earlier in the future. If you are entered for this paper in January, the resources are currently issued at the start of term. If you are sitting in June there is a longer timescale to take account of the half term and the fact that you have other examinations. You should spend at least ten hours of preparation at this stage. Take your time and do not do it all at the last moment.

- Get the teacher to check that you all have the same resources; a missing page is not unheard of when 14 000 copies of a paper are printed. The resource materials are normally contained in a booklet but occasionally there is a separate insert of photographs or even an overlay. The Ordnance Survey map is also separate. It has been known for there to be three packs of resources because of the nature of the issue.
- Do not just go through the resources on your own. Work in a group or with a friend. Do not only work in your groups in the classroom. Arrange some other time outside of the classroom to continue the process. Brainstorm the data.
- Take the materials away and make plans to give time in your revision diary for a thorough examination of the resources. This might be influenced by some class-based activities relating to the resources.
- Learn what the resources actually show and gradually memorise where the information is held. Ask your group or your teacher to explain information that you do not understand.
- If there is an Ordnance Survey map or its equivalent, or photographs, study them thoroughly so that you know what they show. Normally at least one question in

Synoptic unit: issues analysis

- the examination will expect you to demonstrate your knowledge of the map and/or photographs. Make sure you know how to give accurate grid references
- Find out more detail about where the exercise is located in the world. If it is a non-UK exercise, find out about such themes as the climate, the vegetation (including marine life), the landscape forming processes, the level of development of the country, any population characteristics and the nature of the economy. (Some of this may be in the resources.)
- In the light of the location of the exercise and the resources themselves, check with your teacher which parts of your AS and A2 studies are likely to be useful. Reread your notes and textbooks (such as the Heinemann Geography for 16–19 for Edexcel B, *Global Challenges* and *Changing Environments* by Bob Digby, and Dulcie Knifton's *AS Revision Guide for Edexcel B*) on these areas of the syllabus. If there are bits that you do not understand, for example coastal processes or levels of development, ask your teacher for help. This is essential because you are now about to sit an A2 paper when some of your studies of these topics were at AS level and might need updating.
- If there are written resources take note of their origins. The examiners often use comments from individuals. Such comments are normally highly biased because they represent a single viewpoint and can be mistakenly quoted in the examination as fact. Written statements from official bodies and the officers of official organisations probably carry greater weight.
- If there are press reports or media transcripts it is worthwhile researching the political stance of the paper and indeed the nature of the programme from which it was taken.
- Check that you know the big themes and how they might be present or implied in the data. Sustainability of environments, ecosystems, economies and populations is a big theme. Biodiversity, conservation, levels of development, managing change, community participation, and the agencies of change are others.
- If you think that a technique might be useful, check that you know how to do it. Can you draw a matrix, rank proposals, weight factors, sieve maps, undertake a cost benefit analysis, draw a flow diagram showing the critical path through an issue or decision-making process, and draw a bi-polar diagram?
- Are there resources that can be scaled, ranked, weighted?
- If the resource material involves a new location, new activity or moving an activity, is it possible to prepare an Environmental Impact Assessment? Only use EIA if the question demands it or if a question asks you to discuss other techniques that could be used to address an issue.
- Have you got parallel examples of the same issue elsewhere? Sometimes the questions will ask you to relate the issue to similar issues. The purpose is for you to demonstrate your ability to draw on other aspects of your course.
- Because only one copy of the resources is given to you and you have to take that into the examination, get your teacher to obtain further copies from the Edexcel website. Use that copy as a preparation note pad. 'Post It' notes on the diagrams, maps, photos are a useful way of building up your understanding of the resources.
- Go into the examination knowing your stuff and knowing where to find evidence in the resources.

> **Reminder**
>
> Be appreciative of opinions and values expressed on an issue. Always ask why an opinion is held. Many are purely the product of NIMBYism (Not In My Back Yard). This occurs very frequently in a country where people own and value their property and its value might be adversely affected by a landfill site, an airport, an enlarged car park or even mobile phone masts. The same people happily use mobile phones, fly on holiday and park their cars!

> **Reminder**
>
> Do you know the key terms that might be used in the examination? If not check them out and learn them. These can be command words such as 'Describe', 'Explain', 'Justify' and 'Assess'. You also need to know the issues analysis vocabulary of common key ideas that relate to the issue being examined.

At the start of the examination

YOU MUST BRING A CLEAN COPY OF THE RESOURCES WITH YOU TO THE EXAMINATION. If the teacher has run off extra copies and retained the originals, you do not need to worry.

A very good technique at the start of the examination is outlined here. You know what the Resource Booklet and Inserts contain. Draw up a quick Sources Table like the one below as part of your initial planning. **Read the questions and the letter.**

- Use highlighters to note which paragraphs or sentences in the letter refer to each question.
- Note on your table where a figure, or map, or photograph can be used. You now have a guide to all of the relevant information that will help you illustrate your answers.

Synoptic unit: issues analysis

- The final row in the table is the **Synopticity** opportunities and here you can remind yourself of the parts of your AS and A2 course that might be useful.

This example of a Sources Table is based on the June 2003 paper, Breton Bay, West Australia.

Figure No	Question 1	Question 2	Question 3a	Question 3b
LETTER	Para 3	Para 4	Para 5	Para 5
1(a)	*			
1(b)	*			
2(a)	*			
2(b)	*			
2(c)	*		*	
2(d)	*		*	
2(e)	*		*	
3(a)		*	*	*
3(b)*	*			
4		4.2	4.1/4.3	*
5		*	*	
6	*			
7	*			
8		*		*
9		*	*	*
10		*		*
11		*		*
12			*	*
Synopticity	Industry, Population	Coasts, Ecosystems	Industry, Urbanisation	Sustainability of ecosystems, economies

Table 1

- Look at the marks for each question. The total for the question is 60 and there are 10 marks for the Quality of the Written Communication (QWC). Use the following table as a guide to the time that you should take responding to each question. The examination lasts for two hours which means that good timing leads to better marks.
- **Remember to read the letter** again to check that you understand the tasks.
- Use a highlighter to draw attention to the command words such as 'assess', 'to what extent', 'explain' and 'justify' to name but a few. Perhaps use a different highlighter to emphasise the subject matter of the question to stop you drifting to other question topics.

The table gives you the times needed to complete your answers to questions. For example, in a paper with marks as follows Q1 =12, Q2 = 18, Q3 = 18, Q4 = 12 you should expect to spend *5–8 minutes on initial planning*, 20–22 minutes on the first question, 30–33 minutes on each of the next two questions *with 2–3 minutes prior planning* and a final 20–22 minutes *plus 2 minutes planning* on the last question. Planning time can be best used at the outset although some time is needed before you start answering each question. Another way of looking at the marks is to realise that there is 1 mark for every two minutes in the examination room.

Marks	Timing (minutes)
10	15–18
12	20–22
15	25–27
18	30–33
20	35
Planning	10

Table 2 Timing

During the examination

- Keep as strictly as possible to the timings. Do not get bogged down in the first question which is generally for fewer marks.
- Write a very brief plan for each question before you start writing in the answer book.

Synoptic unit: issues analysis

- If you are going to draw up a table, e.g. a conflict matrix or a grid, make sure that the cells are big enough for you to expand your points. Single words convey very little and small grids with one or maybe two words per cell do not score well. You need space for brief points and possibly a reference to the resource that you have used. Always try to construct tables and grids across two pages so that you give yourself the necessary space. Ideally turn the book round and use a double page with no more than four columns across and three cells per page, i.e. five columns with row titles in one, and six rows.
- If you are calculating scores do provide a key saying what the weightings are for factors. Also say what the basis of the scoring actually is. Numbers with no explanation and no key gain virtually no marks.
- Remember that the **10 Quality of Written Communication (QWC)** marks do take account of you answering the question that was set and not one that you think will be the question. The examiner will judge whether your answers are fit for the purpose. So keep to the question!
- Make sure that you spell the words in the question paper correctly. Do not miss-spell the place names; that is careless and will certainly lose you a mark for QWC.
- Make sure you give evidence clearly either by quoting Figure numbers or by making it clear that you are using a source.
- If Ordnance Survey maps are used quote grid references for evidence.
- Do not answer the questions out of sequence, The examination has been set with a definite issues analysis or decision-making sequence in mind. By deviating it is very likely that you will provide the wrong evidence and get questions confused.

Finally, here is a preparation checklist.

	Before resources issued	Once resources issued
Practised mini issues		NA
Sat mock paper		NA
Used AS notes		
Practised with OS maps		
Received ALL resources and made a 2nd copy	NA	
Brainstorm in class and with friends	NA	
Read relevant notes and text books		
Know the terms		
Know where evidence can be found	NA	
Understand the resources	NA	
Have noted similar issues		
Have attempted EIA*		
Have attempted Cost/Benefit*		
Have done scaling, ranking, weighting*		
Can do grid references		

Table 3

NA = Not Applicable; * = Only use in the examination if they are relevant.

A Grade A candidate will be able to:

- support his/her answers with evidence from a wide range of the resources
- analyse data, strengths and weaknesses, points for and against
- write very concisely so that a vast amount of information is conveyed in a short space of time
- write a report that keeps to the sequence required and answers the questions – it is fit for purpose
- use terminology and concepts correctly
- be able to draw on information from the whole of his/her AS/A2 studies
- write good geography.

Answers to exam questions

1.18 (6474, Section A and B)

1 (a) With the aid of examples from a range of economic activities explain what is meant by outsourcing. (10 marks)

Having planned your answer you have 12 minutes to write your response. This is a classic case of the information merely being a stimulus. Describing the scene will gain only one mark.

You should have defined outsourcing. Without that you would not have scored above 7 marks (i.e. the top of the second band). You should have discussed the source regions of outsourcing and the destinations of outsourcing. Outsourcing is not new in manufacturing. Clothing has long been manufactured in LEDCs and NICs. TV assembly was moved to NICs in the 1980s. Today there are two service industry types of outsourcing: (i) back-office activities such as accounts and (ii) call centres. You would need to quote companies, e.g. HSBC; and destinations, e.g. Bangalore. These points logically ordered would gain you 8–10 marks.

(b) Examine the impact of outsourcing on countries at different stages of development. (15 marks)

You have just 18 minutes to answer this section.

The examiner would only expect you to look at two levels of development in the time. You could provide the answer in two columns.

For MEDCs the impacts will primarily be seen as loss of jobs, which in the past brought about the deindustrialisation of textile regions, and today is from the service sector. On occasions the number of losses will be large, e.g. 4000 for HSBC in the UK. Empty premises and perhaps dereliction are further consequences. Greater profits due to lower costs benefit shareholders and investors.

For LEDCs there are the job opportunities that are provided. There are benefits for the employees in terms of quality of life, and for national income as well. There are costs in terms of antisocial hours in the case of call centres, and westernisation and loss of cultural identity. There is the danger of a branch plant economy. If costs rise perhaps the jobs will move elsewhere. This has happened to several electronic branch plants that were established in Penang 20 years ago. They have now been moved again to cheaper Vietnam. These points (not necessarily all of them) would help you reach the top band (i.e. 13–15 marks).

2 (a) Explain the pattern shown on the map. (10 marks)

In this case the map does give you some information that you will need to summarise.

The data shows that HQs are concentrated in North America, the European Union and Japan. There are few elsewhere. The reasons are:

- the strength of the national economies
- the historical position of many of these companies that grew up in colonial times leading to neo-colonialism today
- mergers which have seen the take over of firms as TNCs have grown, e.g. General Motors acquired Opel, and Ford acquired Volvo thus shifting the main HQ to Detroit
- in NICs the attraction of inward investment has enabled many TNCs to strengthen their position, e.g. Siemens in Penang
- protectionism in MEDCs exerted through tariffs that prevent other firms growing
- political domination of the world's one super power
- exceptions could be discussed, such as in Saudi Arabia, but do not focus too much on exceptions to the broad pattern.

Answers to exam questions

(b) With reference to either ONE trans-national manufacturing or ONE trans-national service company, outline and explain the global distribution of its activities. **(15 marks)**

You have a choice here. Name your company straight away and be accurate with your information.

I have added a map below which shows where Bayer, the German chemical group, is located today. Bayer, founded in 1863, was part of the huge I.G. Farben cartel in Nazi Germany that was broken up in 1945. It was one of the key industrial companies in the post-1945 economic miracle in West Germany. Bayer's headquarters are in Leverkusen (the football team was once the works team). Today it employs 122 000 workers: 51 000 in Germany, 19 000 in Europe, 25 000 in the USA, 15 000 in Latin America and 12 000 in Asia and Africa. The company has manufacturing sites that are still mainly focused in Europe (8 plants in Germany) and the USA although there are plants in South Africa and some NICs/RICs but few in Japan because of competition. It is involved in crop science (including GM with Monsanto), health care, chemicals, polymers (fibres). Research is at three locations in Germany, France, the USA and Japan. Administration and marketing are widely distributed and obviously geared to those countries that buy the most of Bayer's products such as Agfa films. That would suffice for a description of its distribution.

Explaining could be a list of the factors. First, there is the strength of Bayer in Germany as a national earner. Research is in Germany but also focused on a handful of G7 countries where there is a strong tradition of advanced research and development in the fields of chemicals and allied activities. There are 12 000 research employees.

Manufacturing is concentrated in the EU, often close to supplies of oil for petrochemicals. Some sites are in development regions to obtain assistance with setting up plants, e.g. Bridgend, UK and Bitterfeld 1994 (Asprin) in Germany. Low cost sites in LEDCs are present, e.g. India and Thailand. Takeovers will account for some production sites, e.g. Cutter Labs in the USA in 1974, and in South Africa.

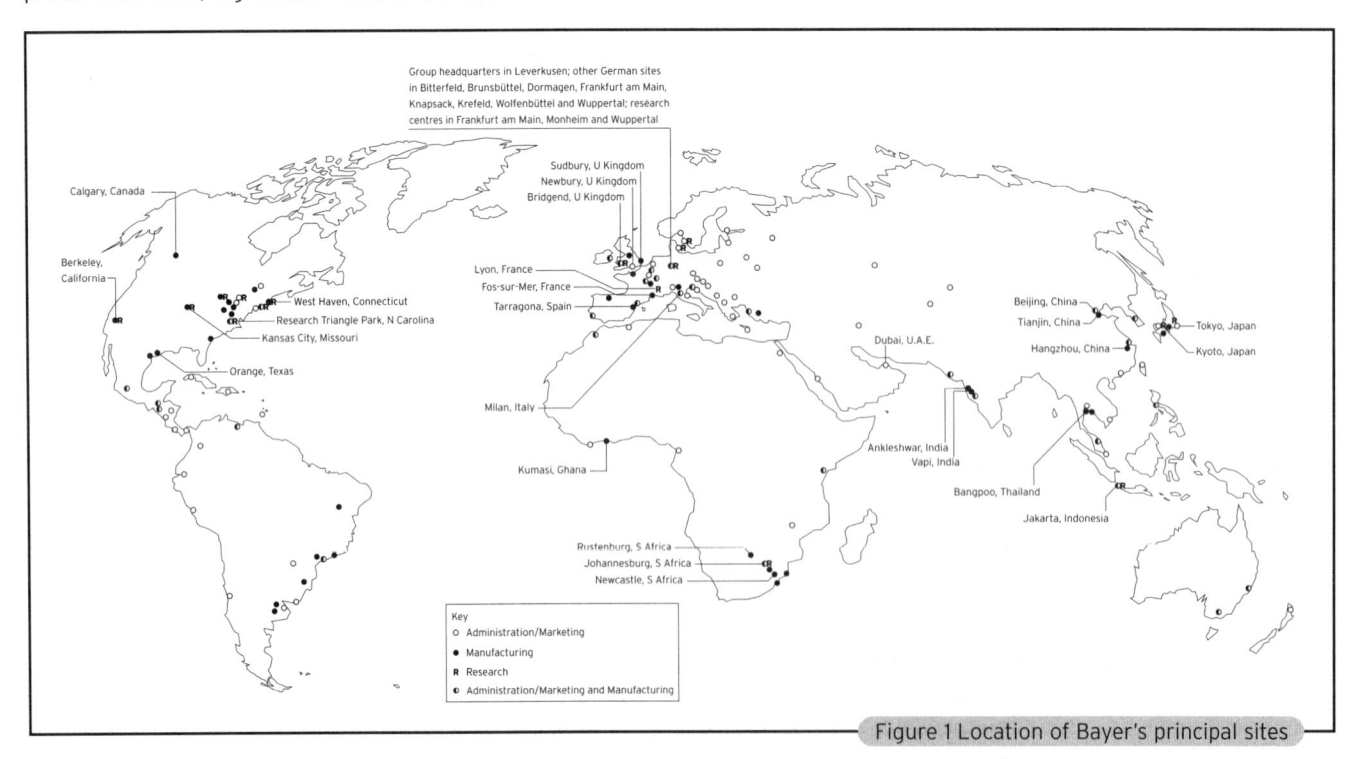

Figure 1 Location of Bayer's principal sites

Administration and Marketing is oriented to areas of demand.

These statements from Bayer publications illustrate other aspects of the company's role in sustainable development.

Answers to exam questions

> Bayer promotes the concept of sustainability as a member of the World Business Council for Sustainable Development (WBCSD), as well as in ECONSENSE, a German industry initiative. The company's goals and activities in this area are published in its Sustainable Development Report.
>
> **Global Compact**
>
> Bayer takes its social responsibility seriously, and was one of the 45 founding members of the Global Compact initiated by UN Secretary General Kofi Annan. Bayer has undertaken to work to the best of its ability to ensure communication and observance of the nine principles selected by the UN in the area of human rights, social standards and the environment. Specific projects supported by Bayer range from improving working conditions for agricultural labourers in South America to supporting the World Health Organisation (WHO) in the fight against sleeping sickness, which threatens 60 million people in Africa.
>
> **Sustainable Development**
>
> Bayer has also pursued the progressive concept of 'in-process environmental protection' since the early 1980s. In line with this concept, new processes and products have been developed to reduce the quantities of emissions and waste forming in the first place. New methods of recycling any residues have supported this trend.

3 (a) Outline the roles of the jet stream and air masses on the weather experienced during the slow north-eastwards passage of the blocking high over the UK. (15 marks)

This question uses the weather maps as a stimulus although they also provide a guide to your answer.

You can divide your answer into two sections. The first section should look at the evolution of the jet stream as outlined in Figure 3 and explain the formation of Rossby waves and how they assist in the creation of a high pressure which is slow moving. You can then go on to explain how the low pressure systems are diverted north and south of the high on the maps. The air masses would be a Tm air mass for the high although a suggestion that it could be TC would not be marked down, particularly since 2003. The air masses further north will be Pm and returning Pm coming in behind the lows. The explanations should be in terms of the characteristics of the high and the effects of its passage NE.

(b) Describe and explain the differences between the effects of blocking highs in summer and winter on NW Europe. (10 marks)

You can go back to the text on page 15 of this book to find the answer. A structured answer in columns or a two part discussion that both describes and then explains can gain very good marks.

1.19 (6474, Section C)

1 (a) Describe and suggest reasons for the relationships between the data shown in the table. (12 marks)

GNP, Infant mortality and Population growth rate are all standard measures of development whereas Total fertility overlaps infant mortality and growth rates in its coverage.

You have got to look at the links between the first column and the other three columns. You will not get more than 4 marks if you just repeat the data either by row or by column. There are LDCs, LEDCs, NICs and MEDC/G8 countries in the data and your answer should try to identify the characteristics of each group. Where there is low GDP, there are high figures in all of the other columns. Singapore bucks the trend with very high GNP and yet it has a higher growth rate. You may comment on the relationship between high GNP and low fertility and growth. All of the NICs and MEDCs have fertility and growth rates that fall below that needed to enable a population to reproduce itself, i.e. 2.3 children per woman.

The reasons are to do with levels of development, levels of education, investment in health and disease control more than birth control in many countries.

Answers to exam questions

If your knowledge of the relationships and your explanations are good, you will get 9–12 marks. If you can note the levels of development but are less able to explain the reasons your marks will probably be between 5 and 8. If you ignore one column of data you would certainly not gain more than 6 marks no matter how good the rest of the answer was. If all you do is 'tour' the data, you will not get more than 4 marks. Low marks will also be awarded if you are highly selective of the data and /or do not offer explanations. Remember the question did say 'suggest reasons'.

(b) To what extent can environmental conditions explain the figures shown in the table? **(18 marks)**

You may refer to both the human and physical environments although your answer should focus on the latter.

You will need to get beyond locating five countries in Africa. Although you may not know all of the geography of the countries, you should be able to comment on environmental conditions such as drought in Somalia and Zambia. Flooding is a problem in the Gambia (and Mozambique recently – do not hesitate to use parallel examples such as this). You will also need to be more explicit than saying these are environments of the North and South. You are expected to note that the poorer countries have a lack of export crops, poor communications and may still suffer from the effects of being a colony in the past (neo-colonialism as a force today).

In the NICs they have overcome the environmental conditions through state-led investment policies. In a similar way Switzerland has overcome environmental conditions or has used them to create wealth-earning activities such as tourism. The educational environment creates a knowledge-based economy in MEDCs.

If you can accurately demonstrate knowledge of some of the environments in the table and are able to assess the role of the environment with support you will score 15–18 marks. If your comprehension of the environment is more limited but your support is patchy and unbalanced either to LEDCs of NICs or MEDCs the marks will probably fall in the range 10–14. Purely descriptive answers without any effort to assess the role of the environment in different countries will gain 5–9 marks. One point or a set of points in note form without any developed reasoning about the role of the environment will only score 1–4 marks.

2 (a) With the aid of examples, discuss how the bomblets interact with one another to cause poverty. **(15 marks)**

You will need to look at several pairings in order to produce a convincing answer. However, it also permits a choice of issue in part (b).

Much will depend on the combinations that you select. You must attempt to link the bomblets and not just write about AIDS without any reference to other bomblets.

One theme is that of population growth without a corresponding growth in resources, food supply leading to famine which in countries with little health care may lead to loss of the wage earners. Environmental abuse as a result of population pressure and debt (cash crops) may lead to famine. You can explore the links between certain types of political power, social equity and low incomes, lack of work etc., e.g. the Taliban in Afghanistan. Lack of timber fuel is in itself abuse of the environment but brought about by population growth and leads to unnecessary labour being expended on fuel wood gathering.

If you write very good linked explanations of the poverty problem supported by well chosen examples and use approximately well over half of the detonators you should score 12–15 marks. Marks will decline if your linkages between the causes are not very explicit. If your supporting examples are increasingly generalized and descriptive rather than explanatory and cover fewer of the detonators, the marks will be between 6 and 11.

Answers which are highly descriptive and about a limited range of bomblets with little support which is exaggerated and often wrong will only gain 1–5 marks.

Answers to exam questions

(b) Select any two bomblets and suggest how these might be tackled to reduce poverty. (15 marks)

These are some ideas which you may include but by no means all. Every point you make must have some support from your studies.

Lack of work might be tackled by social market economy measures as was the case in the 1930s. Government can provide as in Cuba and former USSR, and encourage inward investment, e.g. China.

Bad housing Schemes such as site and service in Zambia. Stopping migration to cities so arresting shanty growth. Housing seen as a right rather than something earned.

Population growth Various forms of population policy – expect one child policy but do expect other policies, e.g. Kerala.

Health care Expect barefoot doctors. State and/or private investment improving health and income potential.

Famine GM crops and Green Revolution. Imports from sales of primary products and industry. Tourism earnings.

Illiteracy Schemes to provide basic education which might be seen as foundation for alleviating other detonators.

Political power How to cope with dictatorships, e.g. Zimbabwe destroying agriculture for political ends. In democracies it might be easier to vote in those with appropriate policies.

Social inequity Again a political issue. Women's rights and the emancipation of women have helped to overcome poverty.

Debt Jubilee 2000 and work of UN on debt relief.

Lack of fuel Alternative fuels such as alcohol from sugar. **Afforestation.** Improved efficiencies of cooking materials.

Low income Chinese policies to provide basic wage. Minimum wage levels. Inward investment by TNCs. Export of primary produce and export-based industries.

Environmental abuse The work of Greenpeace. A host of environmental projects at international level. Company initiatives such as green tourism.

You will get 12–15 marks if you cover two aspects with clear strategies that are discussed in a mature fashion. 6–11 marks will be gained if there is greater variability between the two topic areas selected and one is clear and the other vague. Good understanding of the issues will get you into this band particularly if you have variable support. If your evidence is less strong you will score here. 1–5 marks will be awarded if evidence is patchy and generally descriptive rather than analytical. Strategies that are unclear or simplistic may fall into this marks band.

3 (b) Sustainability is the world's greatest challenge.' Discuss how this challenge may be met in the management of ONE of the following: the atmosphere, the biosphere, population, the economy. (15 marks)

NB This is only the second part of a question. Also remember that you only have to select ONE factor. If you answer more than one the marker will read the whole answer and then mark the best response.

You should have started with a definition of sustainability no matter which of the four options you have selected. Bruntland's 1987 definition is 'Development which meets the needs of the present without compromising the ability of future generations to meet their own needs.' Your introduction should go on to state your option choice. You will also need to introduce the four aspects of sustainability that you revised in 1.9, page 36, i.e. futurity, environment, public participation and equity and social justice.

The Atmosphere
A focus on a particular topic such as Global Warming and enhanced Global Warming would be a good approach. You should examine the international and national attempts to manage global warming. You will have noted the difficulties that politicians

face and how effective citizen participation is difficult because it is so often opposed by industry and the political system. The World Bank, IMF, the UN are some of the other influential organisations. You might discuss how any management might not be equitable because it is unfair to states in the earlier phases of development. On the other hand carbon credits are an attempt to manage equity and social justice.

The Biosphere
Management might be studied by looking at the concept of ecological footprints. Governments and NGOs are aware of the human pressures on the environment whether it is the pressure that results from a highly technological society or a less economically advanced society. The world's footprint is already between 25 and 33 per cent larger than the area available to maintain the world's population.

Management might involve the use of greener energy technologies, e.g. that reduce the use of firewood, reduce CO_2 and therefore the need to expand carbon sinks.

The trading of carbon credits externalise the ecological costs of people's activities by purchasing credit from mainly LEDCs whose footprint is ecologically sustainable. However, this does not reduce pressures on the biosphere and a more radical proposal is to have international agreements that discourage the selling of carbon credits.

The creation of sustainable energy, e.g. wind, wave and solar power, rather than increasing the use of fossil fuels will stabilise the ecological footprint. Some suggest a strategy of green taxes that build in environmental costs into the price of goods and services. National management takes the form of Agenda 21 initiatives at a local level that are required of all local authorities in the UK. Governments can introduce policies that abolish subsidies that create pollution. The management of water is coming to the fore even in countries where water has been regarded as almost a free resource, e.g. the UK in 2003–4 with the threat of drought.

Population
Policies to manage population are more than just the policies to limit children in various parts of the world, e.g. China, Mauritius and Singapore. Sustainable population policies are challenging because the current well-being of populations needs to be maintained and not reduced. People will want choice and some policies could further disadvantage the disadvantaged. The Johannesburg Earth Summit saw poverty as a key force working against well-being. Education of peoples and disease control might feature as management policies at a national level. You could discuss policies to maintain current population numbers in countries where the birth rate has declined and where an ageing society threatens the maintenance of a sustainable population.

People also want community rights and freedom from persecution for any reason, another difficult demand in states where there is a history of cultural differences.

The Economy
No state is close to economic sustainability. There is a basic problem in the economic development of the less developed economies as their development involves activities which threaten environmental sustainability. The most developed economies make excessive demands on the world's ecosystems and will have difficulty maintaining their populations' well-being if they have to restrain economic activity. Much of your discussion might involve examining whether people should use up finite resources at current rates. If development occurs at the current rates in China for example (up to 8 per cent economic growth in 2003) can the resources be found to support that growth? Already China consumes much of the world's genetically modified soya beans and increasing quantities of the plastics, silica and copper needed for modern electronics. Can further growth be sustainable if it takes these resources out of reach of other countries seeking to develop?

Marks in the range of 12–15 will go to those who write a well supported essay in the 20 minutes available. It will have both the introduction suggested above and a full conclusion. Marks in the 8–11 range will go to those who know the principles of management but do not always have examples to support their ideas. These marks could also go to those who have case studies of sustainability but do not relate them to management. 4–7 marks will definitely be the range for those who merely describe a policy such as the China one-child policy without relating it to the question. The lowest band is reserved for short, broadly descriptive statements that are a repeat of knowledge not effectively applied to the question.

4 (b) Discuss the effects that rising population may have on the economy and environment of coastal areas. (18 marks)

NB This is only the second part of a question.

You have the opportunity to select your own case studies when this kind of question is asked. You must provide both environmental and economic case studies.

Environmental

If you have studied marine and coastal ecosystems as your ecosystem option in 1.6–1.8 you should be able to make use of your knowledge of how activities can damage and destroy reefs and other coastal ecosystems. You could make use of UK examples such as the potential effects of developments to serve the population, such as a new port facility at Dibden Bay and its effects on the marshlands in the area. Increased clearance of land by forestry, for settlement and agriculture can result in greater erosion and more sediment transported to the coast. If natural coastal protection, such as mangroves, are removed more erosion may occur. Pollution from industrialisation and relaxed pollution controls, together with the increased volumes of waste piped and dumped in the sea, may cause algae blooms such as those that appear on the Adriatic coast of Italy.

Economic

The growth in population may result in the growth of coastal activities such as fish farming that destroys the natural ecosystems. Increased tourism to many coastal areas of the world, not just the Mediterranean, is the product of an increase in wealthy populations. Tourism results in more developments, perhaps airports on reclaimed land, e.g. Chep Lap Kok, Hong Kong (not just a tourist airport), and migration of hotel workers and travel representatives to the resorts in search of work.

Population growth often results in new ports and harbours to cope with trade growth and the seaward expansion of docks and installations, e.g. Singapore, and Europort. There is an economic multiplier effect which will inevitably attract service activities for the growing population.

Coasts are socially attractive as retirement areas, e.g. Sussex and Devon or West Wales. Proximity to the coast is often used to attract skilled work forces, e.g. Sophia Antipolis Science Park near Nice, insurance firms to Portsmouth, Poole and Bournemouth and IBMs HQ relocation from London to Portsmouth in the 1970s.

The top marks of 14–18 will be gained if you have a broad set of effects that balance both economic and environmental. You will need to select your examples from a range of countries at different stages of development. Good answers will demonstrate that not all development has a negative effect.

Answers that are less broad and probably focus too strongly on two or three examples will gain 9–13 marks. If you start to limit your essay to just economic or just the environment you will not score above 8 or 9 marks. If your essay lacks support and is very generalised description 4–8 marks will be gained. Vague effects without evidence will score in this band. If the effects are all seen as negative or if your answer is very negative and sensationalist with gloom and doom statements it will only score up to 3 marks. If it is note form and only a few lines and if it lacks any examples, it could also find itself with a mark below 3.

Index

Page numbers in italics refer to illustrations.

aid 56-7
air masses 12
   UK 13, *13*
anticyclones 15
avalanches 83, 89-90

biodiversity hotspots 37-8, *38*
biomes 24, *24*
   forest 25
   grassland 25
   marine 26

climate
   see also weather
   extremes 10, 81-2
   hazard responses 90
conservation 36-7
coral reefs 34-5, *34*

demographic replacement 41
demographic transition models 42, *42*
depressions 15
   mid-latitude 14, *14*, 15
droughts 83, 87-8

earthquakes 82-3, 84, *84*
economic groupings, international 52-3
ecosystems 24
   green issues awareness 69-70
   human impact 26, *26*
   importance 26-7
   threats to 27
ecotourism 70
El Niño 18-19, *19*

female rights, LEDCs 45
foreign direct investment (FDI) 61-2
forests
   biomes 25
   boreal 30, *30*
   energy flows *28*
   nutrient cycling 28, *28*
   tropical rain- 29-30, *29*
frontogenesis 12

global warming
   effects 21, *22*
   greenhouse effect 20-1, *20*
   solutions 21-3
globalisation 57
   technological transfers 64-5
grasslands
   desertification 32-3, *33*
   temperate 31-2, *32*

   tropical 31, *31*
greenhouse effect 20
   enhanced 20-1
   natural *20*

hazards
   perceptions of 91
   predictions 92-3, *92*
   responses 90
hurricane predictions *92*

immigration, UK 51, *51*
InterTropical Convergence Zone (ITCZ) 16, *17*

jet streams 12, *12*
   Polar Front 14, *14*

Kondratieff Waves 58
Kyoto protocol 1997 22, 106, *107*

La Niña 19, *19*
least developed countries (LDC) 53

mangrove swamps 34-5
migrations
   causes 47
   forced 47
   laws of 48-9
   refugees 49-51, *50*
   voluntary 48
Millenium Summit 2000 37
monsoon climates 16, *17*, 18

Niña, La 19, *19*
Niño, El 18-19, *19*
north-south divide 53-4, *54*

ozone layer 23

pollution
   and economic development 96-7
   economic impacts 100-1, *101*, *102*
   environmental impacts 99
   human causes 103
   increases 97-8, *97*
   Kyoto protocol 22, 106, *107*
   local initiatives 107-8
   management of 104-5
   monitoring 94-5, 104
   national initiatives 107
   natural causes 103
   scales of 96
   social impacts 100, *101*, *102*
   studying 94

populations
   age structures 45, *45*, *46*
   changes *40*, 44
   densities *39*, *40*
   growth *42*, 43-4
   growth theories 43-4
poverty alleviation 67-8
   NGO role 68, *68*
product life cycles 60, *79*

rainforests, tropical 29-30, *29*
refugees 49-51, *50*
resources
   and economic development 78-9
   exploitation management 79-80
   finite 78
   renewable 78
   wildernesses 113-14
Rossby waves 12

seasonality, climate 16
service industries
   growth 58-60, *58*
   outsourcing 62-3, *63*
sustainability gap 69

technological transfers 64-5
tornadoes 81-2, 84-5, *85*
tourism, ecological 70
trade
   global economy 55-6
   globalisation 57
trans-national companies (TNC) 61

UN Development Plan 66-8
UN Environmental Programme 37

volcanoes 83, 86-7

weather
   see also climate
   deterministic predictions 10, *11*
   deterministic responses 10, *11*
   forecasts 9-10
   technological fix 10, *11*
wilderness environments
   classifications 109-10
   definitions 110
   indigenous populations 111-12
   international agreements 115
   national boundary problems 117-18
   national management 116
   pressures on 113-14
   protection of 111, 117-18
   threats to 112